AIRCRAFT TECH KNOWLEDGE

for the Private Pilot

Book 6
Aviation Books Series

DR STEPHEN WALMSLEY

Copyright © 2022 Walmsley Publications

All rights reserved

No part of this book may be reproduced, stored in a retrieval system, or transmitted in any form or by any means, electronic, mechanical, photocopy, recording, or otherwise, without express written permission of the publisher.

Disclaimer

Whilst every effort has been made to ensure the accuracy of the information, the author does not give any guarantee as to its accuracy or otherwise. Nothing in the contents of this book is to be interpreted as constituting instruction or advice relating to practical flying. Students preparing for their relevant exam should consult training their syllabus published by the relevant authority to ensure they are studying towards the most up to date syllabus. The author shall not be liable nor responsible to any person or entity concerning errors and omission, or loss or damage caused directly or indirectly by the use of the information contained in this book.

ISBN (paperback): 9798437926543
ISBD (hardcover): 9798438470779

AVIATION BOOKS SERIES

The aviation books series provides the reader with an educational and enjoyable reading experience. A focus has been placed on practical, hands-on aviation by linking science with the real world.

Private Pilot series includes:

Book 1: Human Factors

Book 2: Aviation Weather

Book 3: Flight Radio

Book 4: Principles of Flight

Book 5: Flight Navigation

Book 6: Aircraft Technical Knowledge

Book 7: Flight Instruments

Commercial Pilot series includes:

Book 8: Pilot Performance & Limitations

Book 9: Meteorology

Book 10: Aerodynamics

CONTENTS

AIRCRAFT TECHNICAL KNOWLEDGE
Copyright
Aviation Books Series
Preface
Introduction 1
Chapter 1: Airframe Structure 4
Chapter 2: Control Systems 14
Chapter 3: Undercarriage 25
Chapter 4: Aircraft Engine 30
Chapter 5: Exhaust System 41
Chapter 6: Induction System 45
Chapter 7: Ignition System 56
Chapter 8: Oil and Cooling Systems 61
Chapter 9: Propellers 71
Chapter 10: Engine Management 83
Chapter 11: Fuel 91
Chapter 12: Electrical System 101
Conclusion 108
Index 110
Books in this series 112
About The Author 115
Commercial Pilot Series 117

PREFACE

Aircraft are amazing machines that can leap into the sky with seemingly little effort. However, aircraft are complicated, and pilots must understand them to fly safely. Understanding the theory side of the aircraft is essential for all pilots, giving them the confidence to make the right decision when things are not running smoothly.

The focus of this book is on airframe structures, engines and systems in single-engine, propeller-driven aircraft. This book follows closely the syllabi of *Aircraft Technical Knowledge* from a range of aviation authorities around the world. This book goes beyond these syllabi, with a particular focus on practical aviation, linking science with the real world. Each chapter contains a range of visual figures in full color and mini case studies that will allow the reader to have a deeper understanding of the aircraft.

INTRODUCTION

Aircraft are amazing machines that can leap into the sky with seemingly little effort. The basic design principles of today's aircraft have not changed an awful lot since the early days of flight. Aircraft are built strong to withstand the considerable loads placed on them throughout a flight but still remain relatively lightweight. A powerful engine is essential to generate sufficient thrust to allow the aircraft to fly, and the engine must be reliable. Hidden throughout the aircraft are various supporting systems; from the electrical system to the engine cooling system, all doing their part to help the aircraft safely reach its destination. Aircraft technical knowledge is one of the fundamental topics a pilot must master to operate an aircraft safely. This topic is not just about understanding the different components of the aircraft, but also recognizing when something is not right.

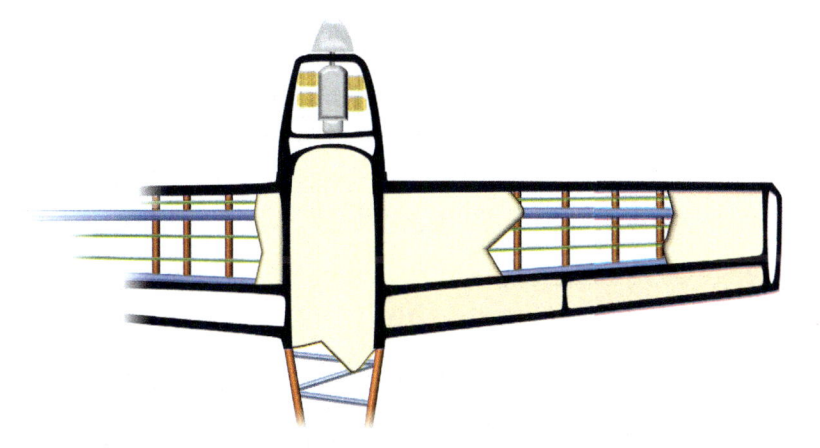

You will start your aircraft technical knowledge journey by exploring the major airframe components. Although aircraft are designed for a variety of purposes, from training to aerobatics, most have the same major airframe components. These include the fuselage, wing and empennage (tail section). You will see these have evolved to keep up with the higher loads placed on them as aircraft have become faster and larger. This includes developments in the basic building blocks, with composite materials becoming more popular in modern small aircraft. Next, you will explore the flight controls and undercarriage (landing gear). The flight controls are relatively small compared to the main airframe components, but without them, the aircraft would not be able to maneuver in the air. The undercarriage will go unnoticed throughout most of the flight, but as the aircraft touches down on the runway, it becomes one of the most important parts of the aircraft. You will then look at the aircraft engine and its various supporting systems. An aircraft engine is able to convert fuel into mechanical energy that turns the propeller. But it cannot do this task alone, relying on various systems to help deliver the correct fuel to the engine and take away the enormous amount of heat it generates. Despite aircraft engines being reliable, occasionally they'll have an issue, such as running rough, or the oil temperature rises at an alarming rate. Understanding the engine will help pilots troubleshoot when things are not running smoothly. The final chapter explores the aircraft electrical system. If the electrical system is not treated properly it can malfunction, which can be very dangerous.

Knowledge of the aircraft will help you treat the aircraft properly, from smoothly operating the throttle to making sure the aircraft is using the correct type of oil. If an aircraft is treated poorly, the consequences can be severe. The focus of this book is on how general aviation aircraft fly – primarily single-engine, propeller-driven aircraft. However, many of the design principles of small aircraft can be applied to aircraft of all sizes. No matter what aircraft you fly, it is essential you have a solid understanding of aircraft technical knowledge.

CHAPTER 1: **AIRFRAME STRUCTURE**

If you gazed out across a busy aerodrome, you may notice aircraft of a range of shapes and sizes. The wonderful thing about aviation is aircraft of a considerable age remain actively flying, which means it is possible to view how airframe structures have advanced through the years. You can view currently flying aircraft that have a wooden frame, covered in cloth, through to aircraft with a full composite structure. An airframe structure is an amazing piece of engineering – each part, from the fuselage, wing and tail section, is incredibly strong. However, at the same time, the structure needs to remain as light as possible to allow the aircraft to leap into the air.

The airframe structure can be split up into several main components, as shown in Figure 1.1:

- The **fuselage** is the main body of the aircraft.
- The **wings** (also called mainplanes) are the main lift producer.
- The **empennage** is sometimes called the aircraft's tail.
- The **undercarriage, engine** and **control surfaces** (e.g. elevator, rudder, aileron) are also structures and will be explored in the next few chapters.

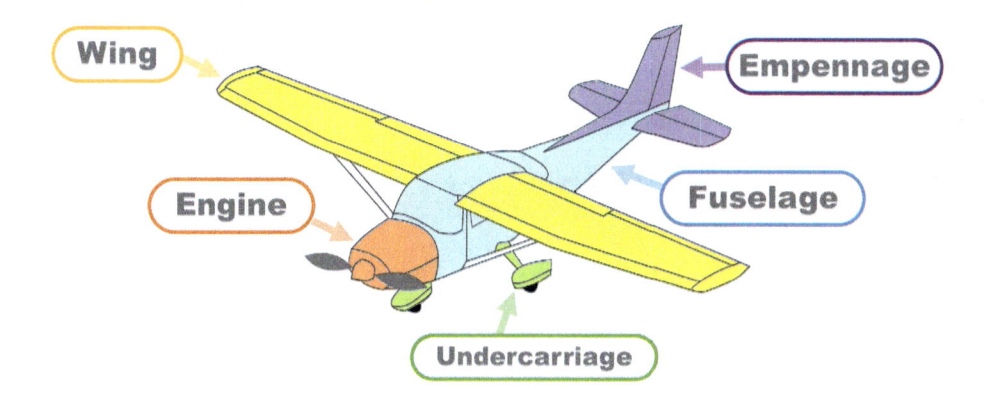

Figure 1.1: *Main components of an aircraft.*

Each airframe structure needs to be strong enough to withstand the considerable bending and twisting loads (forces) during a flight. If the structure is not strong enough, it will break or deform. Over the years airframe structures have gone through several design changes that have largely come about due to aircraft becoming larger and faster, therefore being exposed to greater loads.

Fuselage

Let's begin by exploring the body of the aircraft, also known as the fuselage, which has a number of key functions. Firstly, the other key components of the aircraft are attached to the fuselage, including the tail section, wings, engine and undercarriage. Secondly, the fuselage contains the cabin in which the pilot, passengers and cargo are located.

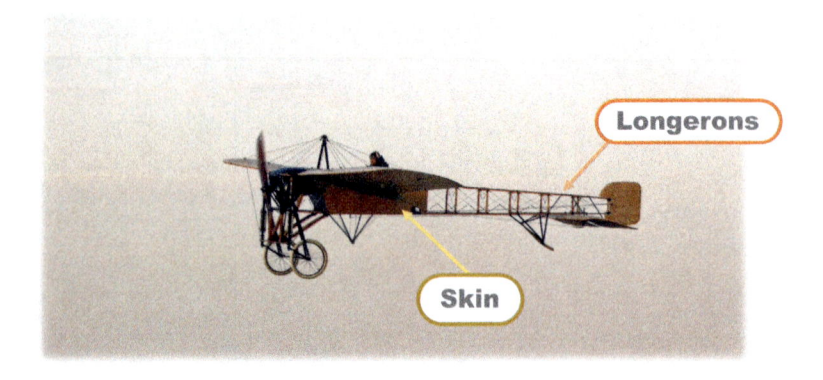

Figure 1.2: *Earliest fuselage designs looked like a box-like structure.*

The earliest fuselage design is known as a **truss-type** construction. Essentially this design is a wooden or metal box-like frame, with a skin covering (typically fabric covering), as shown in Figure 1.2. In this design, the frame supports the majority of the load, with the skin providing some support. The frame is made up of several longerons, struts and braces, as shown in Figure 1.3. **Longerons** are the long frame lengths that run from the nose to the tail, whereas **struts** and braces are attached to the longerons (both vertically and horizontally) to share the load placed on the frame. Stringers or formers are also sometimes used to provide shape to the fuselage (more on these items in the next section).

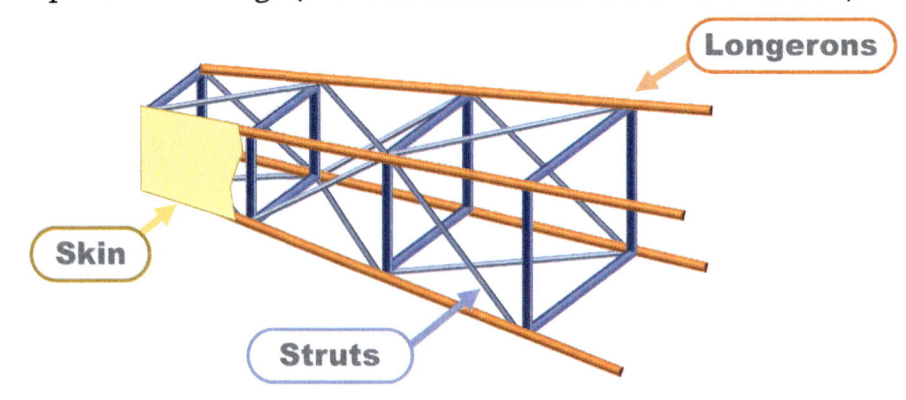

Figure 1.3: *Typical components in a truss-type fuselage.*

The overall result is a box-like structure that works fine for old aircraft that traveled slowly, but as aircraft became faster, streamlining became very important. Streamlining basically means how easily the air can flow around an object. When air struggles to remain in a streamlined flow, it can result in a considerable increase in the drag on the aircraft – drag is the force that resists the movement of the aircraft through the air. Today, truss-type designs are only used in the construction of some small aircraft. As technology improved, aircraft have been designed to allow the skin to support some or all of the flight load placed on the fuselage, especially with the replacement of fabric skin with lightweight metals, such as aluminum. This has led to the monocoque and semi-monocoque designs, which are more likely to be used in modern aircraft.

Monocoque means single shell. In this fuselage design, the external skin (or shell) supports all the load, with no internal framing. Everyday examples include an egg or an aluminum beverage can. It is rare to find a pure monocoque structure in an aircraft (i.e. no framework at all), but rather aircraft with a monocoque design usually have some internal framing, such as formers or bulkheads, which are used to attach the skin and help provide its basic shape, as shown in Figure 1.4. A monocoque design can be very strong, but not in all directions. For example, an aluminum beverage can is able to support a large force at the ends of the can, but not so much from the side, which can easily deform and collapse. Other issues of this design include the skin usually needs to be thick, meaning more weight, and there are construction complications in cutting hatches or doors into the skin of the aircraft fuselage (which degrades the shell's strength). As a result, most modern aircraft are a mix of the two designs discussed (truss and monocoque), known as a semi-monocoque design.

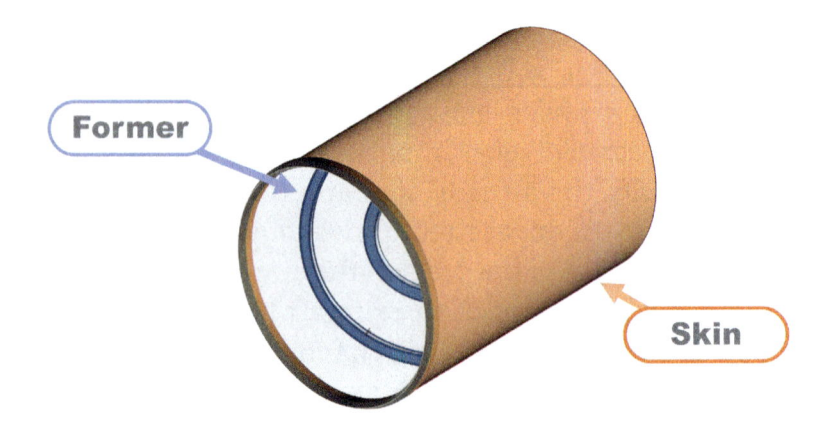

Figure 1.4: *Example of a monocoque structure.*

A **semi-monocoque** fuselage structure is the most common design found in aircraft, large and small. Loads are shared by both the skin and the frame. The frame generally consists of stringers and formers, as shown in Figure 1.5. Stringers are much smaller and lighter than longerons (which are used in the truss-type design), with the formers providing shape, similar to the monocoque design. The skin – often a lightweight aluminum – is attached to the frame. Collectively, this design is better at withstanding the bending and twisting loads that an aircraft is exposed to during a flight. Furthermore, this design is more tolerant to skin damage, as the other parts of the structure can share the load.

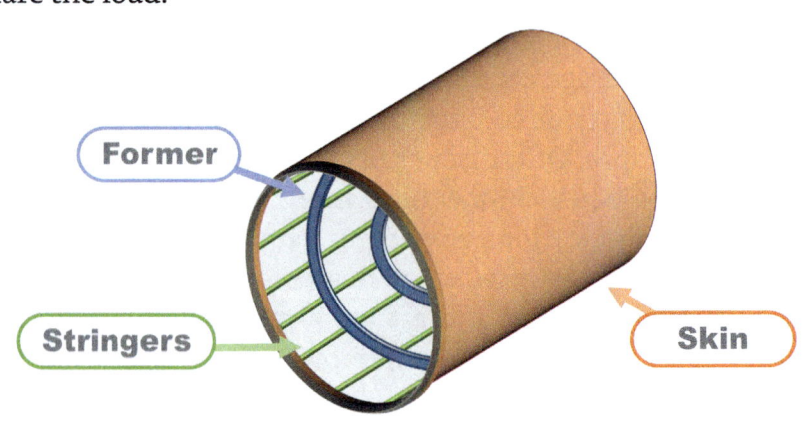

Figure 1.5: *Typical semi-monocoque fuselage structure.*

Composite

Traditionally, airframe structures have been constructed with metal frames and aluminum skin, materials that are still heavily used in small aircraft today. However, composite materials are slowly making their way into aircraft structures, including the fuselage. Composite is a broad term that covers a wide range of materials, including fiberglass and carbon fiber (and even within these, there are several subgroups). Compared to traditional materials such as metal, composite materials in airframe structures have many advantages, including being lightweight, strong and corrosion resistant. A typical composite material consists of bundles of glass or carbon fibers that are woven into a matrix system or cloth. This is then 'glued' together with a resin that hardens. Usually, several bundles or layers are used. A major advantage of composite materials is that during this manufacturing process they can be placed in or over a mold, allowing specific shapes to be formed. Aircraft have been using composite materials for some time, especially for non-load bearing structures, such as fairings. However, there is an increasing trend to use composite materials for larger load-bearing structures, such as wings and fuselage sections.

One of the main disadvantages of operating an aircraft with composite material is it can be very difficult to recognize damage to the structure. When a pilot conducts their preflight inspection of the aircraft, one of the many items they should be looking out for is damage to the skin or airframe structure, such as from a bird strike on a previous flight or a tool dropped on the aircraft when it was in the hanger. Damage to an aircraft with aluminum skin can often be easily recognized in the form of a dent. If a dent is visible, the pilot can see straight away that the aircraft will need to be properly inspected to determine if it is safe to operate. However, damage to composite material is not always visible on the surface, especially following a low energy impact like a tool being dropped on the structure. Despite no visible surface damage, underneath the impact area there may be delamination, which is internal damage to the composite material. As such, even minor impacts should be properly inspected to determine if underlying damage exists. Other signs of damage to composite material include a dull or whiteish color compared to surrounding areas. Composite material is still generally more expensive than traditional material, which is one of the reasons modern aircraft still heavily use aluminum skin and frames.

Wing

The wing is one of the most important structures of an aircraft. Due to the special shape of the wing, it is able to generate sufficient lift to overcome the weight of the aircraft, allowing it to leap into the air. The wings need to be able to withstand loads well in excess of the aircraft's weight, which can place considerable stress on the wing. During a flight, the wing loading is usually upwards, but when the aircraft lands, the wings tend to droop downwards under their own weight. There are several different wing configurations, each designed for different types of operations. Most aircraft only have a single wing, known as a **monoplane,** but some have two sets of wings and are called **biplanes,** as shown in Figure 1.6. Modern biplane designs are more likely to be found on aerobatic aircraft.

Figure 1.6: *Examples of different wing configurations.*

Wings can be attached to the top, middle or lower portion of the fuselage, as shown in Figure 1.7. To help share the load between the wing and the fuselage, some aircraft may have external braces (especially high wing aircraft). This is also known as a braced or **semi-cantilever** wing type. If a wing has no external braces – which is common for a low wing aircraft – it is known as a **full cantilever** wing design. This means the external bracing is not required to carry loads between the wing and the fuselage.

Figure 1.7: *High and low wing configurations, including bracing.*

The wing has several key structural components, as shown in Figure 1.8:

> The main load-bearing structural component of the wing is the **spar**, which runs the length of the wing. Wings will often have several spars, even on smaller aircraft. A common design is a main spar that runs along the length of the wing at the point of maximum thickness, and a smaller, lighter spar towards the rear of the wing to which the flaps and ailerons are attached.

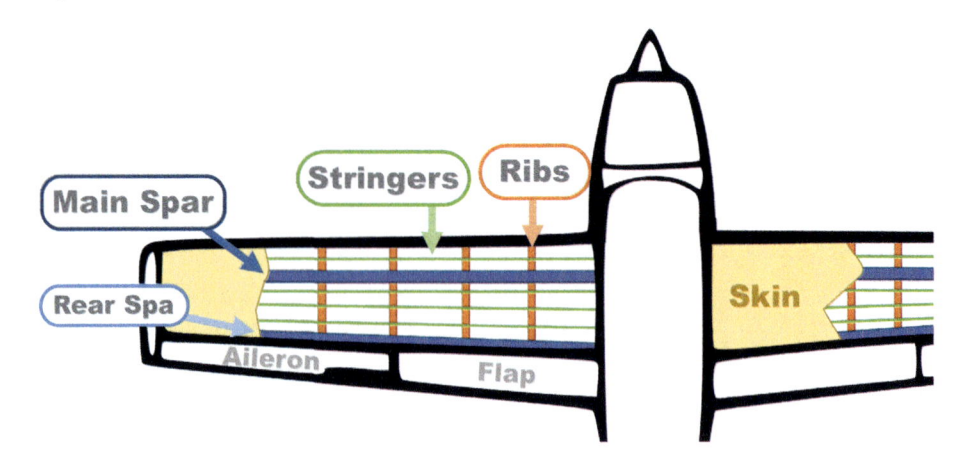

Figure 1.8: *Typical structural components of a wing.*

Attached to the spar will be several **ribs**, which run perpendicular to the spar and are very important as they provide the aerofoil shape of the wing. The aerofoil shape is one of the key features of the wing that help generate lift, generally comprising of a curved upper surface, and flatter lower surface. The aerofoil shape of the wing varies depending on the specific purpose of the aircraft. For example, a thin wing aerofoil is generally better for high-speed flight, whereas a thicker wing aerofoil is better suited for small aircraft that travel slower.

The **skin** will support some loading (known as a stressed skin type) and is usually a thin metal, like aluminum alloy.

Running in the same direction as the spar are several **stringers,** to which the skin is also attached, helping to share the loading.

In most modern aircraft, the fuel tanks are also integrated into the wing structure (more on fuel tanks in chapter 11).

Empennage

The final airframe structure to explore here is found at the very back of the aircraft and is known as the empennage, more commonly called the tail section. This includes the vertical stabilizer (fin) to which the rudder is connected and the horizontal stabilizer that the elevator is connected to, as shown in Figure 1.9. Some aircraft may have an all-moving horizontal stabilizer (i.e. no separate elevator), which will be explored in the next chapter. The empennage normally has a similar construction to the wing, whereby a number of spars and ribs are used, along with stressed skin to help share and absorb the load.

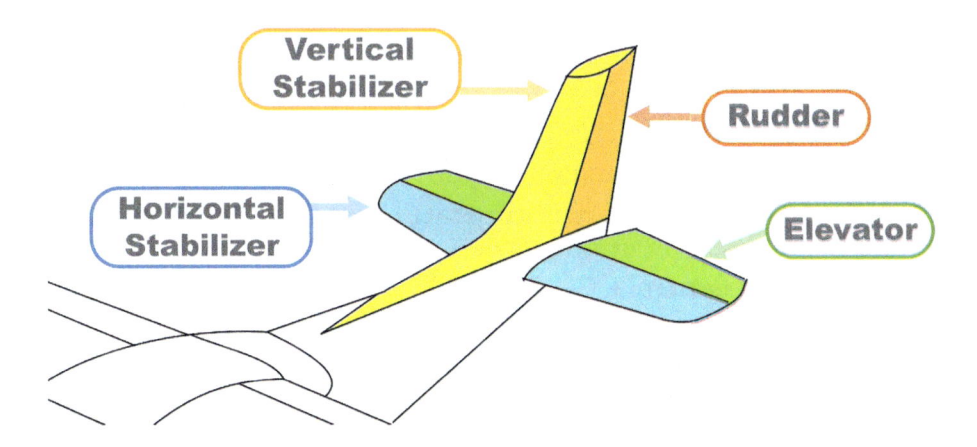

Figure 1.9: *Typical components found in the aircraft empennage (tail section).*

Modern aircraft may follow different design principles; from low to high wings, from aluminum to composite material. Despite these differences, they all have a common goal; to provide a strong, lightweight structure that can withstand the considerable and varying loads placed on them during a flight. Next, you will explore some of the smallest aircraft components – flight controls, which can have a large influence on an aircraft's ability to fly.

CHAPTER 2: **CONTROL SYSTEMS**

You explored the large components of the aircraft in the previous chapter, such as the fuselage and wing, which are designed to be strong and lightweight. But during a flight you will need to maneuver the aircraft; from pitching the nose up in a climb to rolling the wings in a turn. This is achieved using the aircraft's control systems, which use small control surfaces that adjust the lift and drag at various points around the aircraft. There are also control systems that are designed to improve the performance of the aircraft and relieve excessive pressure on the control column. In this chapter, you will explore each of these control systems and how pilots operate them from the cockpit.

Primary Flight Controls

The aircraft rotates around three axes, which are three imaginary lines that pass through the aircraft's center of gravity (CG), as shown in Figure 2.1. The CG can be considered a point at which all weight of the aircraft is concentrated.

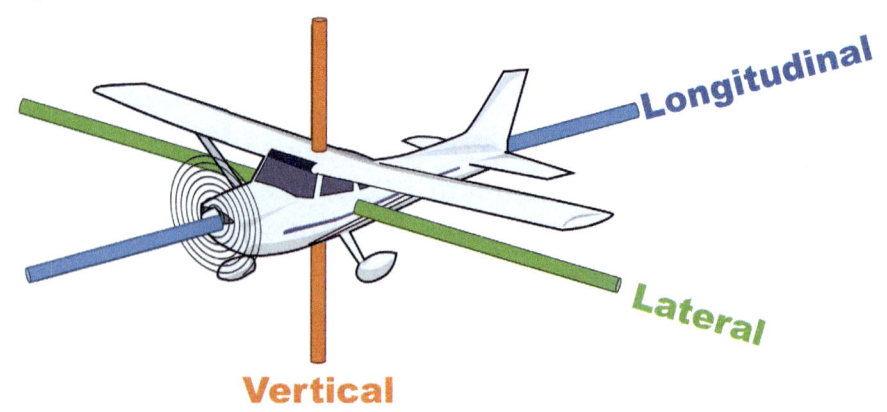

Figure 2.1: *Aircraft axes.*

- The axis that runs from the nose to the tail of the aircraft is known as the **longitudinal axis**. The aircraft **rolls** around the longitudinal axis.
- The axis that runs from wingtip to wingtip is known as the **lateral axis** and the aircraft **pitches** around this axis.
- The axis that is at right angles to the other two is known as the **vertical axis** and the aircraft **yaws** around this axis. This axis may also be referred to as the 'normal' axis.

The three motions of the aircraft (pitch, roll, yaw) are controlled by three types of control surfaces, as shown in Figure 2.2. Pitch is controlled by the elevator, roll by the ailerons, and yaw by the rudder. These are known as primary flight controls and are required to control an aircraft safely during a flight.

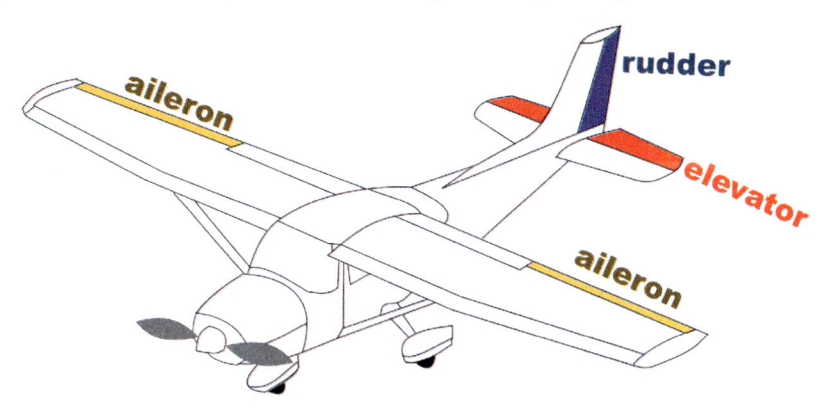

Figure 2.2: *The aircraft's primary flight controls.*

Ailerons control the roll of the aircraft around the longitudinal axis. Pilots move the ailerons by rotating the control column left or right (a control column may also be called a yoke, control wheel or control stick). The ailerons are attached to the outboard section of the aircraft's wing and move in the opposite direction to each other. For example, consider an aircraft that is rolling to the right, as shown in Figure 2.3. When the pilot moves the control column, the left aileron will move downwards whereas the right aileron moves upwards, which generates greater lift on the left wing, allowing the aircraft to roll to the right. The aileron's cross-section normally continues the same profile as the wing, which means when the ailerons are in the neutral position, the aircraft will not roll.

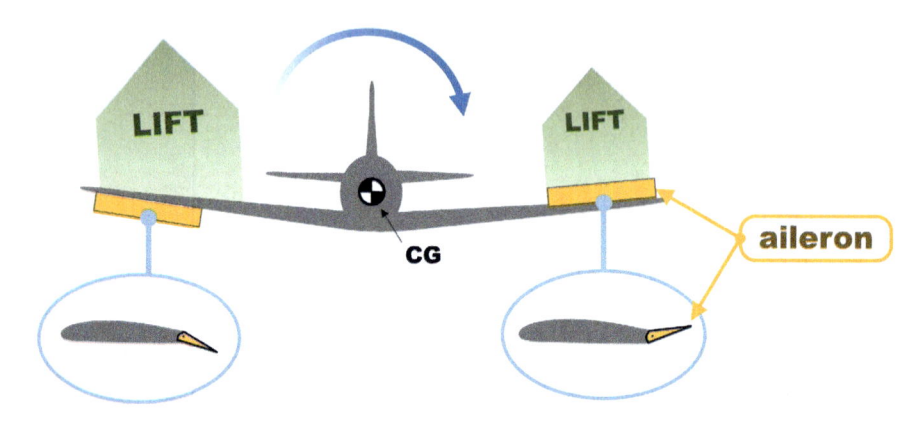

Figure 2.3: *Ailerons deflect in opposite directions, allowing the aircraft to roll around the center of gravity (CG).*

The **elevator** controls the pitch of the aircraft around the lateral axis, which means moving the nose of the aircraft up or down. The pilot moves the control column forward and back to operate the elevator. When the pilot wishes to pitch the nose up or down, lift is changed on the elevator. The elevator is attached to the end of the horizontal tailplane (also known as a conventional tailplane). A second type of design to control pitch is known as an all-moving tailplane (also called a stabilator) whereby the whole horizontal tail surface moves around a central pivot point, as shown in Figure 2.4. In terms of pitching, they operate in a similar way (adjusting the lift on the horizontal tail surface).

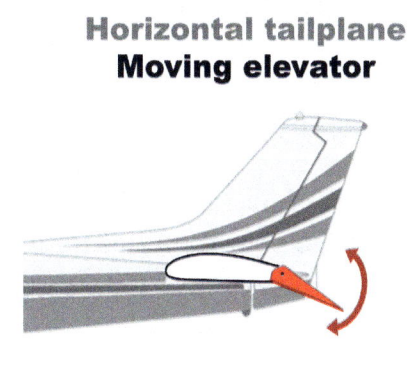

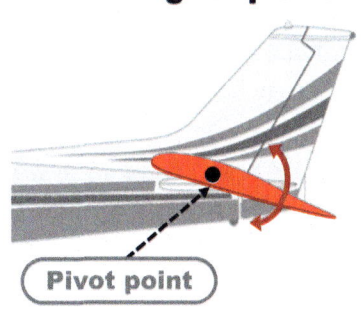

Figure 2.4: *An elevator is attached to the horizontal tailplane, whereas the all-moving tailplane (stabilator) moves around a pivot point.*

The **rudder** controls the yaw movement of the aircraft around the vertical axis. Yaw basically means the aircraft's nose is not pointing in the same direction as the flight path. For example, the aircraft in Figure 2.5 is traveling up the page, even though the nose of the aircraft is pointing towards the left (yawing left). The rudder is a moveable control surface fixed to the vertical stabilizer or fin and is controlled by the rudder pedals in the cockpit.

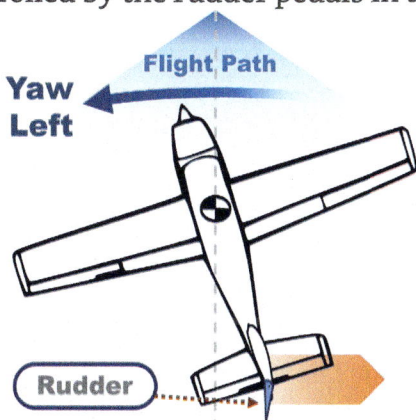

Figure 2.5: *The rudder creates a sideways force, allowing the aircraft to yaw.*

Control surfaces are usually constructed similar to the parent surface it is attached to. For example, the ailerons will have a similar construction to the wing, with spars, ribs and skin. However, as they are smaller, they can be made with a lighter weight construction. In small aircraft, the primary flight controls are connected to the cockpit by a series of cables and pullies (and sometimes pushrods), as shown in Figure 2.6. This is known as a mechanical or manual flight control system. The pilot does not need any extra assistance to move the control surfaces. The feel on the control column – known as stick force – will be determined by the deflection (movement) of the control surface and the speed of the aircraft. In larger or faster aircraft, extra assistance may be required to move the controls, similar to the power steering in a car, but this is not normally required in small aircraft.

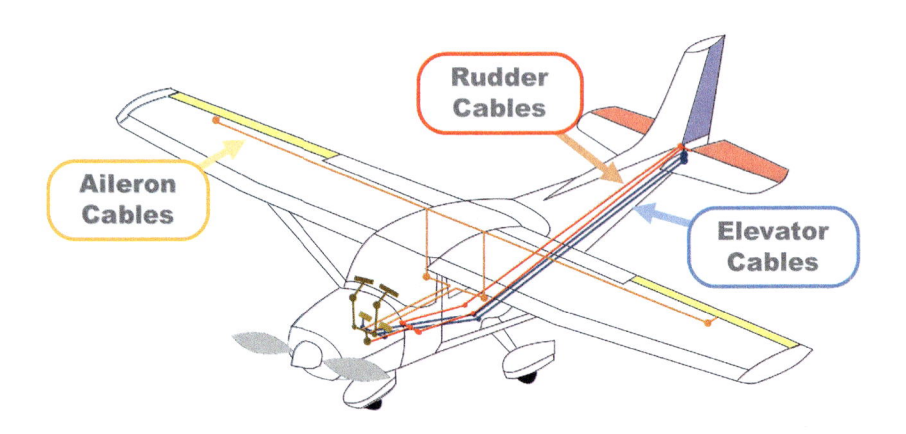

Figure 2.6: *Cable-operated controls that may be found in a small aircraft (note: the actual cable location will vary depending on the aircraft design).*

Each primary flight control should be carefully inspected before a flight. This inspection is not only looking for damage (e.g. bending or dents) but also checking that each control moves freely and in the correct direction – the importance of this is shown in the following case study:

> On the 23rd of July 2017, a pilot was preparing for a flight in San Jose, California in a single-engine Piper PA28. The aircraft had undergone extensive maintenance prior to the flight, including the installation of a new rudder pedal and brake control. Shortly after take-off the aircraft immediately entered an uncontrolled left roll. The pilot was unable to control the aircraft, resulting in the aircraft crashing beside the runway. During the post-accident investigation, it was found the ailerons were working in the opposite sense. When the control column was moved to the right, the right aileron moved down and the left aileron moved up, which would have rolled the aircraft to the left (the opposite direction to the control column movement). The aileron cable was disconnected for the maintenance before the flight (to allow access to the rudder cable) but was reconnected backward. Prior to take-off, the pilot checked to make sure the controls were moving freely, however did not verify they were moving in the correct direction. This case highlights the importance of not only checking the free movement of the flight controls but also that they are moving in the correct direction – especially after an aircraft has undergone maintenance.

Some of the primary controls will also have a control lock. A control lock is designed to protect a control surface on the ground during a windy day. When the aircraft is parked, a gust of wind can catch the control surface, potentially damaging it and other parts of the aircraft. On small aircraft, the control lock for the elevator and ailerons is usually set in the cockpit, with either a switch or device that holds the control column in a set position. The rudder control lock may be connected to the rudder itself and must be removed during the preflight inspection. It is very important that any control lock is removed before flight, and this is often a checklist item that the pilot should follow (including checking the controls move freely and in the correct direction). If the control lock is accidentally left on, the consequences can be serious, as shown in the following case study:

> In 2014, a Gulfstream business jet with seven people on board raced down the runway for take-off. The experienced pilots were unable to rotate the aircraft to leap into the air, resulting in the aircraft crashing at the end of the runway. The reason for failing to rotate? The control lock was still engaged. The aircraft's checklist required the pilots to confirm the free movement of the flight controls before take-off, but the pilots failed to run through the checklist. Investigators found the captain of the flight had tried to memorize the checklist. The accident report noted that executing checklists by memory removes many of their benefits and leaves a crew more susceptible to errors.

Secondary Flight Controls

Even small aircraft have a range of secondary controls, which are designed to improve the aircraft's performance and relieve excessive loading on the control column. The most common secondary flight control is the flap, which offers the best of both worlds, extra lift at low speed, allowing the aircraft to leap off the runway quicker, but no drag penalty at high speed, once the flaps are retracted. One of the smallest secondary controls can offer the greatest benefit to the pilot – the trim tab. This small device is attached to the end of some control surfaces, making it easier to fly the aircraft (see later for more details).

Flaps

Flaps are located on the inner trailing edge of the wing, as shown in Figure 2.7. They are operated from the cockpit, either with a lever or an electrical switch. Most flaps can be extended (lowered) in stages. Generally, a small flap setting is used during take-off (more lift for small drag penalty), whereas a larger flap setting is used on landing (full flap extension), as the larger amount of drag helps fly a steeper approach and lowers the aircraft's landing speed.

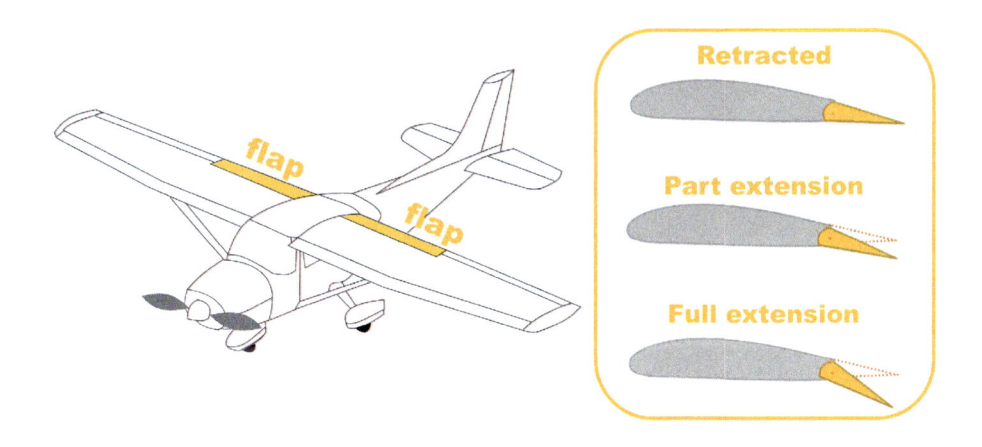

Figure 2.7: *Flaps are located on the inner trailing edge of the wing and can usually be extended (lowered) in stages.*

There are four main types of flap: plain, split, slotted and Fowler flap.

> The **plain flap** is the simplest of the four flaps. As shown in Figure 2.8, the plain flap is hinged to the back of the wing and pivots down when extended. The plain flap increases lift but also greatly increases drag.

The **split flap** is deflected from the lower surface of the wing and generates slightly more lift than the plain flap, but also creates more drag. When fully extended, both the plain and split flap produce a high amount of drag for very little extra lift.

The most common type of flap found on small aircraft is the **slotted flap**. The slotted flap significantly increases lift compared to a plain or split flap. The slotted flap is similar to the plain flap, however, it has a gap between the flap and the wing. Air from below the wing flows through the slot and then over the upper surface of the flap, improving the lifting capability of the wing.

The **Fowler flap** is a type of slotted flap, which also increases the wing area. Instead of rotating on a hinge, the Fowler flap slides backwards on tracks. The first stage of a Fowler flap will increase lift significantly for very little drag penalty.

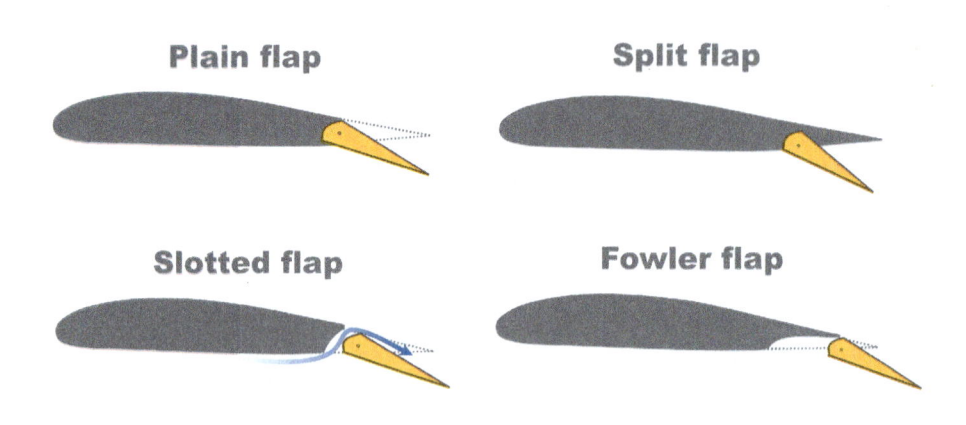

Figure 2.8: *The four main types of flaps (plain, split, slotted and Fowler).*

Trim Tab

A trim tab is designed to hold a control surface at a set deflection, which relieves the pilot from having to constantly hold pressure on the control column (e.g. hold the elevator in a position for a steady climb). Trim tabs may be found on all three primary flight controls (elevator, aileron, rudder), however, in small aircraft they tend to be limited to the elevator (pitch) and rudder (yaw). The trim tab moves in the opposite direction to the primary control surface, creating a force in the opposite direction to the main force, replacing the pressure the pilot must exert on the control column to keep the attitude stable. For example, imagine you are holding forward pressure on the control column during a descent, meaning the main elevator control surface will be in the down position, as shown in Figure 2.9. To keep the aircraft in this attitude, the pilot can either hold the control column in a forward position or have a trim tab do the work for them. In Figure 2.9, the trim tab moves up (in the opposite direction to the elevator), therefore holding the elevator in the desired position.

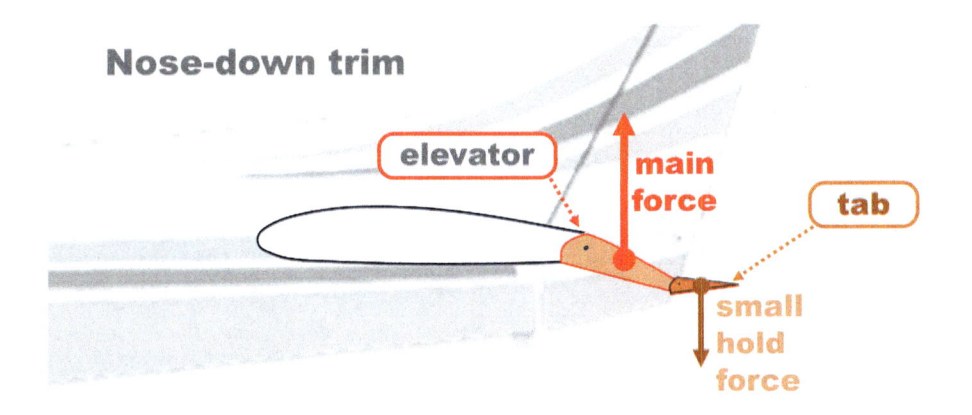

Figure 2.9: *A trim tab moves in the opposite direction to the main control surface, to relieve the constant pressure on the controls.*

Despite being small, a trim tab is a very powerful device that reduces a pilot's workload. Most trim systems are adjustable, meaning the trim setting can be changed from the cockpit. The normal trim procedure is: first set the desired attitude, and then trim in the desired direction until the aircraft is balanced. If the power setting or speed changes, the trim setting will need to be reset.

Control systems are designed to give the pilot sufficient responsiveness to control inputs while allowing a natural feel. These flight controls give the pilot the ability to operate the aircraft as desired; from rolling the aircraft onto the runway centerline to pitching up after take-off. Primary and secondary flight controls can vary considerably between different aircraft types; therefore, it is essential pilots familiarize themselves with the control systems in the aircraft they are flying.

CHAPTER 3: **UNDERCARRIAGE**

The final airframe structure to explore is the undercarriage, also known as the landing gear. During a flight, the undercarriage can be a burden, as most small aircraft have a fixed undercarriage, which means it cannot be retracted in flight. Even with the aid of fairings to streamline the airflow, the undercarriage can be a considerable source of drag as it hangs beneath the aircraft. However, as the aircraft touches down on the runway, the undercarriage suddenly becomes the most important structure by supporting the weight of the aircraft on the ground. The undercarriage needs to not only support the aircraft's weight, but also be strong enough to withstand the considerable vertical load that is placed on it during touch down, and sideways loads that may be experienced during a crosswind landing. But the job of the undercarriage does not finish on touchdown, as it must also help the pilot steer the aircraft and bring it to a stop, not easy tasks when the aircraft is traveling at high speeds.

Most small aircraft have an undercarriage with three wheels, two wheels located below the wing area that carry most of the load, and a third wheel either in front or behind, as shown in Figure 3.1. An aircraft with the third wheel located behind the main wheels is usually referred to as a tailwheel aircraft or **taildragger.** An aircraft with the wheel at the front is known as a nosewheel aircraft or aircraft with **tricycle gear.**

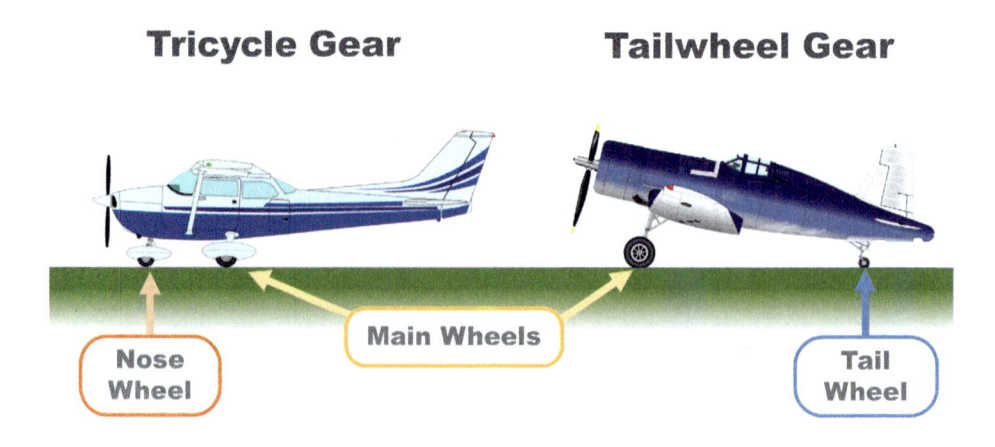

Figure 3.1: *Most small aircraft have an undercarriage with three wheels, two wheels located below the wing area and a third wheel either in front or behind.*

A tricycle undercarriage design is the most common type found on small aircraft as they are:

- Easier to handle on the ground.
- Have less tendency to 'nose over' during braking at high speed (e.g. on landing).
- Have better forward visibility during ground operations, including taxi, take-off and landing.
- Tend to be more stable during take-off and landing, especially during a crosswind.

The main advantage of a tailwheel undercarriage is a larger propeller may be fitted to the aircraft (which helps produce more thrust) and they are better suited for unpaved surfaces (e.g. gravel runways).

The actual undercarriage itself comes in a number of designs, with the shock absorption system and oleo strut the most common types found on small aircraft, as shown in Figure 3.2:

A leg **shock absorption system**, such as a tubular spring steel design, is sometimes used for the main undercarriage. This is a simple design, generally made of strong strips of metal that absorb and dissipate the landing load.

The **oleo-pneumatic shock** strut is used in aircraft of all sizes. This system uses air and oil in two separate cylinders, with the air expanding or compressing depending on the load placed on it. The oil helps to dampen rebound action placed on the undercarriage, such as during a firm landing where a bounce may occur. The greater the load on the strut (e.g. on landing), the more the air is compressed. Many aircraft use the oleo-strut on all three wheels, whereas others may only have it on the nosewheel (like the aircraft in Figure 3.2).

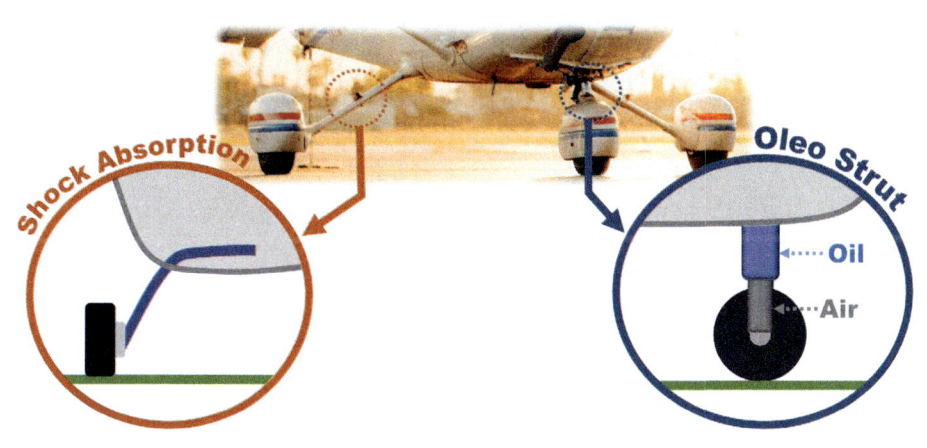

Figure 3.2: *Examples of a shock absorption and oleo strut undercarriage design.*

Aircraft with a tricycle undercarriage will usually have nosewheel steering, whereby the nosewheel is connected to the rudder pedals through control rods or cables – the same rudder pedals used to control the yaw of the aircraft during a flight, as shown in Figure 3.3. Other aircraft may have a castering nosewheel, which is not connected to the rudder pedals and is free to move. If the aircraft is traveling fast enough the rudder pedals can still be used for steering (by steering with the aircraft's rudder) or through differential braking of the main wheels (e.g. when the brake on the left main wheel is applied, the aircraft turns to the left). Most tailwheel aircraft have a castering tailwheel, and therefore steer using the rudder (when traveling fast enough) or through differential braking.

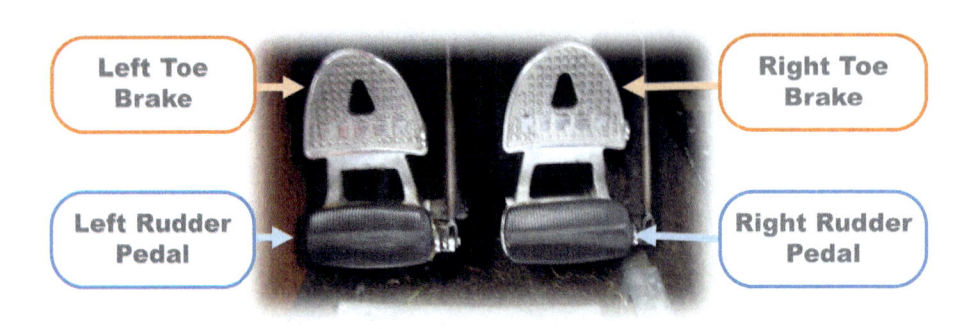

Figure 3.3: *Example of the rudder pedals and brakes of a small aircraft (with nose wheel steering).*

Most small aircraft have disc brakes located with the two main wheels, and these are hydraulicly operated. Aircraft usually have toe brake pedals, which are located just above the normal rudder pedals, as shown in Figure 3.3. Both brakes operate independently, therefore to brake straight ahead (e.g. on landing), both pedals need to be applied with the same amount of force.

During most flights, the undercarriage will go unnoticed, but if it malfunctions, it will certainly get the pilot's attention. This is why a careful preflight inspection of the undercarriage is essential. Keep an eye out for any signs of damage on the undercarriage struts, including leaking oil from an oleo-strut. The tire condition should be carefully checked, including inspecting for proper inflation, tread depth and wear. Some tires may also have creep marks, as shown in Figure 3.4. Creep occurs when the tire rotates around the rim, which can occur when the stationary wheel is forced to rotate very fast on touch down. If the tire moves around the rim too much, it can compromise the seal between the rim and the tire. During the preflight inspection pilots should check that the two creep marks are aligned, if not, the tire is at risk of deflating.

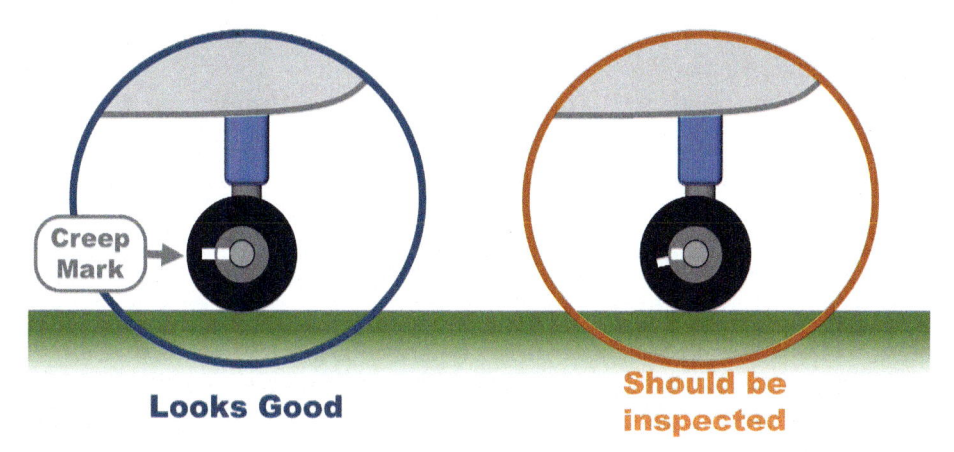

Figure 3.4: *Creep marks should be checked during a preflight inspection.*

You have now explored the main airframe structures, from the fuselage to the undercarriage. Next, you will take a closer look at the key aircraft component that helps generate thrust – the engine.

CHAPTER 4: **AIRCRAFT ENGINE**

Although the engines found in small aircraft come in many shapes and sizes, they all have one very important characteristic – reliability. Modern engines are very reliable, with a major engine failure a rare event. There is a good reason for this, as when an aircraft engine stops in flight, the consequences can be very serious. The most common type of engine found in small aircraft is known as a piston or reciprocating engine. With the help of the propeller, the engine produces thrust to allow an aircraft to fly. Collectively the engine and propeller may also be referred to as the powerplant. For the engine to operate smoothly, it also needs a range of support systems, such as ignition, cooling and exhaust systems, which will be covered in subsequent chapters.

The piston engine works on the basic principle of taking chemical energy (fuel) and converting it into mechanical energy. This conversion occurs within several cylinders in the engine, where a mixture of fuel and air is ignited and rapidly expands, driving down a piston inside the cylinder. There are two main types of piston engines; **spark ignition** and **compression ignition** engines. We will initially focus on the spark ignition engine which is more likely to be found in small aircraft.

Basic Components

To start with, let's explore the key components of a typical piston engine, as shown in Figure 4.1.

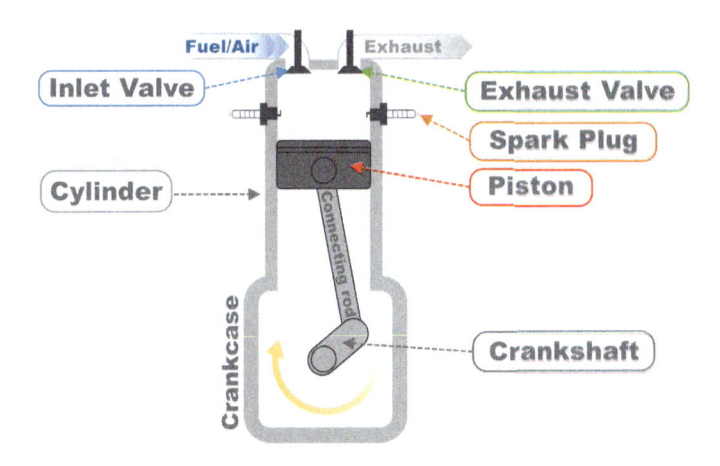

Figure 4.1: *Basic parts of a piston (reciprocating) engine.*

- The **cylinder** provides an enclosed place where the fuel-air mixture is compressed and burned.
- A **piston** is found inside each cylinder and moves back and forth within the cylinder.
- A **connecting rod** connects the piston to the **crankshaft**. In most small aircraft the crankshaft directly connects to the propeller. Collectively they convert the back-and-forth motion of the piston into radial motion, allowing the propeller to turn.
- A spark ignition engine will usually have two **spark plugs** near the top (head) of each cylinder.
- **Inlet and exhaust valves** are located at the top of the cylinder and allow the fuel-air mixture to enter the cylinder, and the exhaust gases to exit.

Cylinder Layout

An aircraft engine requires multiple cylinders to produce enough power for flight, with small aircraft usually having four or six cylinders. These cylinders are arranged around the crankshaft, which connects to the propeller. Several different cylinder arrangements have been used in aircraft engines, as shown in Figure 4.2.

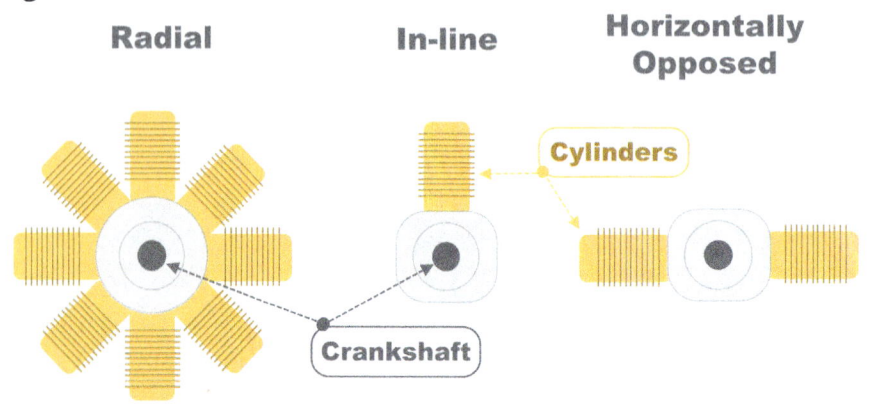

Figure 4.2: *Examples of cylinder layouts (looking face on).*

Older engines often used a **radial** arrangement, whereby a row or rows of cylinders are arranged in a circular pattern around the crankshaft. These engines have a reasonable power-to-weight ratio and good cooling ability (as each cylinder has good access to the cooler outside air – known as air-cooling). However, one of the downsides of the radial arrangement is it has a large frontal area (area hitting the incoming airflow). This is great for cooling but makes streamlining the aircraft difficult (therefore producing higher drag). Some engines have been developed with an **in-line** cylinder arrangement (one cylinder behind the other), which has a smaller frontal area, making streamlining easier. The cylinders can be arranged either above the crankshaft like the example in Figure 4.2, known as in-line upright, or below, known as in-line inverted. The in-line cylinder arrangement tends to have a low power-to-weight ratio, and the rearmost cylinders have poor cooling (due to limited access to cooler outside air).

Figure 4.3: *Example of a horizontally-opposed engine (with the propeller detached).*

The **horizontally-opposed** engine is the most common cylinder arrangement found in small aircraft, as shown in Figure 4.3. These engines have an even number of cylinders (e.g. four, six), and each cylinder will have an opposed or opposite cylinder. This arrangement still has a relatively small frontal area allowing good streamlining but also has sufficient air cooling.

Basic Principles

Power is produced when a fuel-air mixture is ignited inside each cylinder, rapidly expanding and pushing the piston, which turns the crankshaft (via the connecting rod). You will note it's a fuel-air *mixture* that is ignited, not just fuel. Fuel needs oxygen to burn and therefore before arriving at the cylinder fuel is mixed with air - this will be explored in more detail in chapter six. The piston moves back and forth (or up and down), known as a stroke (moving up is one stroke, moving down is another). Most piston engines operate on a **four-stroke cycle**. Nikolaus Otto developed the four-stroke engine; and this cycle is also known as the *Otto cycle*. The four strokes in the cycle are known as the intake, compression, power and exhaust stroke, as shown in Figure 4.4:

Intake Stroke: The piston is moving downwards, and the intake valve at the top of the cylinder is open. This causes the fuel-air mixture to enter the cylinder. This stage may also be called the suction stroke, as when the piston moves downwards, it creates low pressure inside the cylinder, causing the fuel-air mixture to be sucked in.

Compression Stroke: The intake valve is closed and the piston is moving upwards. The fuel-air mixture is compressed, helping produce a greater amount of power once the fuel-air mixture is ignited. Compressing also causes the temperature of the fuel-air mixture to rise.

Power Stroke: The fuel-air mixture is ignited by the spark plugs, causing a huge increase in pressure as the gases expand, forcing the piston downwards. This is the power that turns the crankshaft. This stage may also be called the expansion stroke.

Exhaust Stroke: The burnt fuel-air mixture turns into exhaust gases that need to be removed from the cylinder. In this stroke, the exhaust valve opens and the piston moves upwards, pushing the exhaust gases out of the cylinder.

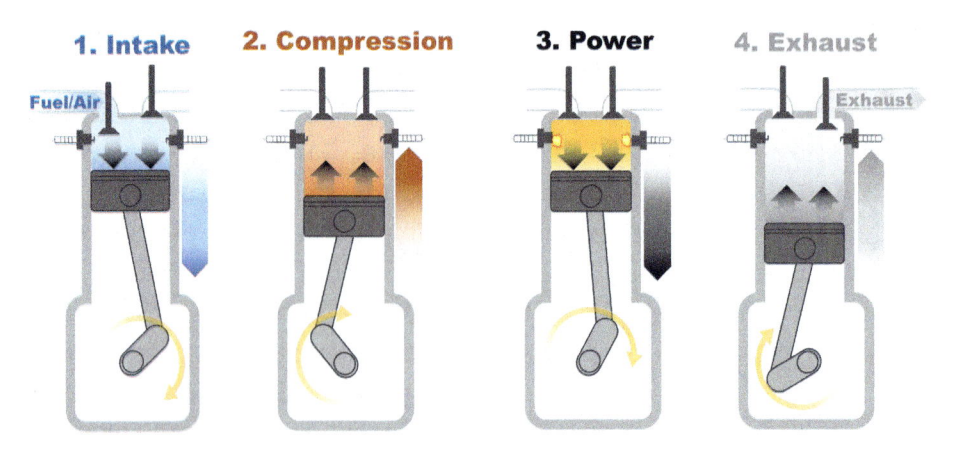

Figure 4.4: *The four-stroke engine cycle.*

During the four-stroke cycle, only one of the strokes develops power, but the crankshaft has rotated twice. To ensure the engine runs smoothly, engines with multiple cylinders are designed to ensure the power stroke occurs at different times. For example, when one cylinder is on an intake stroke (sucking in the fuel-air mixture), another cylinder may be on a power stroke. This works particularly well with a four-cylinder engine, which is common in small aircraft, with all four cylinders on a different stroke at any time.

Valve Timing

The typical engine speed of a small aircraft in the cruise is about 2,400 revolutions per minute (rpm). Each valve will open and close once every two revolutions of the crankshaft, meaning each valve opens and closes a staggering 1,200 times per minute or 20 times a second.

In its most basic configuration, the exhaust valve (that lets out exhaust gases) opens at the beginning of the exhaust stroke when the piston is at bottom-dead-center (BDC) and closes as the piston reaches top-dead-center (TDC). The intake valve (that lets in the fuel-air mixture) opens at the beginning of the intake stroke, TDC, and closes as the piston reaches BDC. However, to obtain maximum power from the engine, the timing of the intake valve and the exhaust valve is usually adjusted slightly. The main goal is to squeeze in as much fuel-air mixture as possible, which can be converted into useful power:

> *Valve Overlap:* Let's start at the end of the exhaust stroke, whereby the piston is moving upwards, pushing out the exhaust gases through the open exhaust valve. To help get the maximum amount of fuel-air mixture into the cylinder, the intake valve is opened a little *before* the piston reaches the top of the stroke (TDC) as shown on the top left of Figure

4.5. This means both valves are open at the same time – fuel-air mixture entering the cylinder, and exhaust gases exiting. The exhaust valve will close slightly after the piston starts heading back down during the intake stroke.

Valve Lag: As the piston begins the compression stroke, the intake valve remains open for a short time as the piston heads upwards, known as valve lag, as shown on the bottom left of Figure 4.5. This allows a little extra fuel-air mixture to enter the cylinder.

Valve Lead: Both valves remain closed during the power stroke, however, before the piston reaches the bottom position (BDC), the exhaust valve opens, known as valve lead, allowing the exhaust gases to start exiting before the piston begins the exhaust stroke.

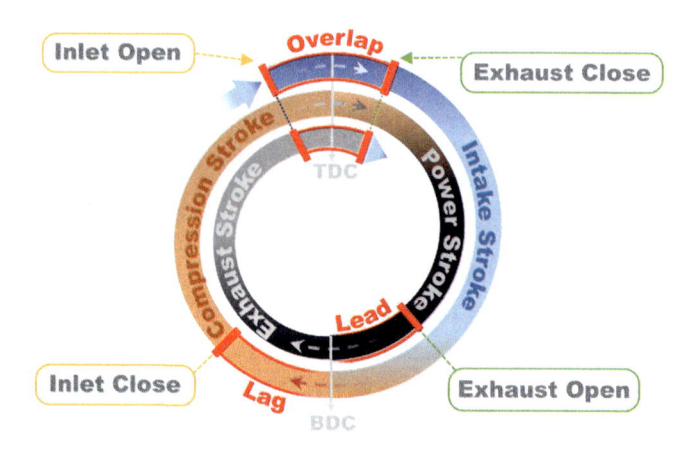

Figure 4.5: *Typical timing of the valves in a four-stroke cycle.*

Detonation and Pre-ignition

During normal combustion, the fuel-air mixture will burn in a controlled manner, which means a gradual build-up of temperature and pressure within the cylinder. This ensures the piston is smoothly pushed downwards during the power stroke. However, if the fuel-air mixture burns in an uncontrolled manner, more like an explosion, it can lead to **detonation,** as shown in Figure 4.6. Detonation can cause excessive temperatures and pressures in the cylinder, which may result in the failure of parts of the engine (e.g. piston, cylinder), leading to a loss of engine power (or complete engine failure). Detonation may be caused by using the wrong fuel (more on fuel types in a later chapter) or operating a very hot engine (e.g. climbing for long periods at a high power setting and low airspeed).

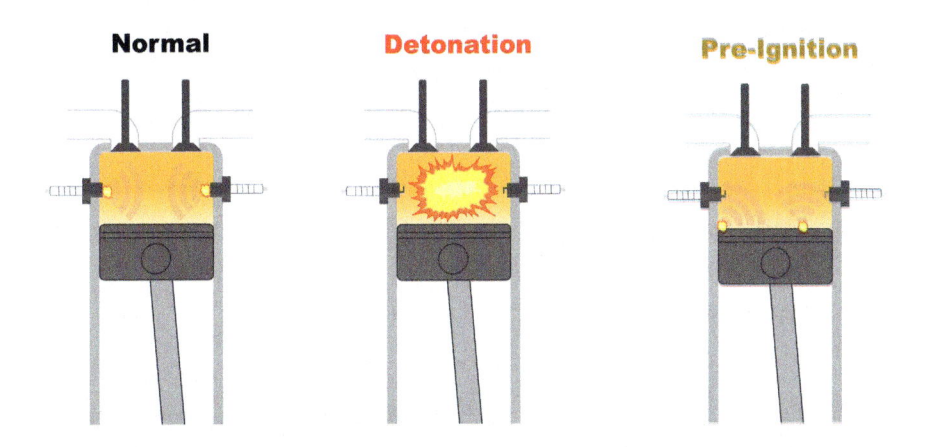

Figure 4.6: *Detonation and pre-ignition can damage the engine, leading to an engine failure.*

Pre-ignition means the fuel-air mixture ignites *before* the normal ignition timing. This is usually caused by hot spots in the cylinder, such as on the spark plug or a damaged part of the cylinder. Like detonation, loss of power may be experienced. It is not uncommon to have detonation and pre-ignition occurring at the

same time, as both may occur when the engine is very hot. If pre-ignition or detonation is suspected, you can consider:
- reducing power
- increase the richness of the fuel-air mixture (more fuel in the mixture helps cool the cylinders)
- increasing airspeed (to aid with air-cooling).

You will explore some of these items in more detail in chapter ten (*engine management*).

Compression Ignition (Diesel) Engine

The piston engine that we have just explored will usually run on AVGAS (aviation gasoline), and the fuel-air mixture is ignited by spark plugs. This engine design has served small aircraft very well, resulting in reliable operations with good power output. However, the compression ignition piston engine is starting to be used in increasing numbers in small aircraft. They may be better known as diesel engines, as they run on diesel or jet fuel (jet fuel is the same fuel used in turbine-powered aircraft).

The main components are the same as a spark ignition piston engine; cylinders, pistons, valves, connecting rods and a crankshaft, and usually they operate with a four-stroke cycle. The key differences are highlighted in the stages of the four-stroke cycle below:

> *Intake Stroke:* As the piston moves downwards, **only air** enters the cylinder. This is different from the spark ignition engine, which has a fuel-air mixture entering the cylinder at this stage.
>
> *Compression Stroke:* The piston moves upwards and compresses the air. The air is compressed to very high pressure, resulting in the air becoming very hot.
>
> *Power Stroke:* Once the piston reaches the top of the stroke, fuel is injected into the cylinder and mixed into the hot

compressed air, as shown in Figure 4.7. As the air is very hot, the fuel spontaneously ignites, forcing the piston downwards (hence, there is no need for spark plugs).

Exhaust Stroke: The exhaust gases are then removed through the exhaust valve.

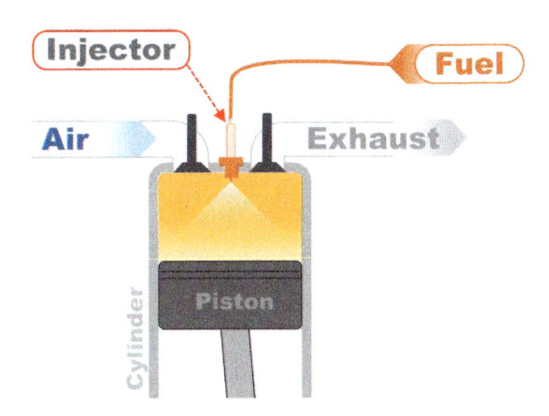

Figure 4.7: *A compression ignition engine ignites the fuel by injecting fuel into the very hot compressed air in the cylinder.*

Diesel engines have been used in cars for a long time but their uptake in aviation has been slow. However, this is gradually changing as jet fuel is often more accessible and cheaper, and the engine usually consumes less fuel (because it is more efficient).

Two-stroke Engine

The four-stroke engine remains the most common design found in small aircraft engines, however, modern two-stroke engines are starting to appear (although mainly in very light aircraft, such as ultralights). As the name indicates, the whole cycle is completed in two strokes, rather than four. A number of two-stroke designs have been used, with a basic example shown in Figure 4.8:

Stroke 1 **Intake, Compression and Exhaust Stroke**: The fuel-air mixture is forced into the cylinder via an inlet port. At the same time exhaust gases exit through the exhaust port located near the top of the cylinder. As the piston moves upwards, it compresses the incoming mixture.

Stroke 2 **Power Stroke:** The piston closes off both the exhaust and fuel-air mixture ports. The mixture is then ignited, pushing the piston downwards.

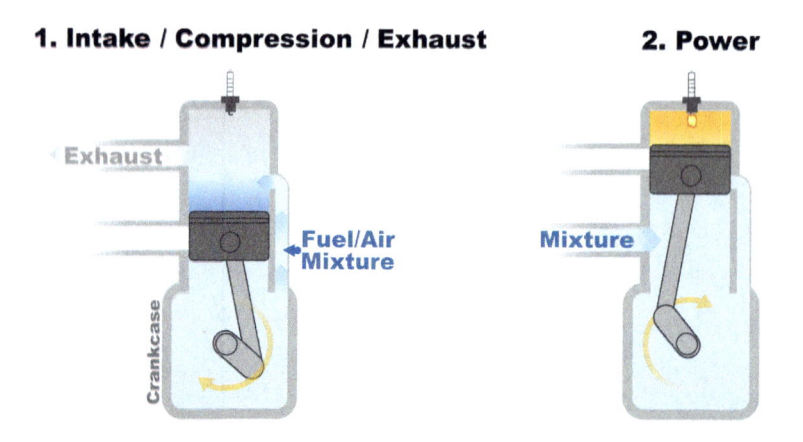

Figure 4.8: *Basic example of a two-stroke cycle.*

Traditionally, two-stroke engines were not suitable in aircraft engines, but newer designs are showing promise and they provide a compact engine with fewer moving parts. Two-stroke engines have also been developed using compression ignition.

Engines on small aircraft are an amazing piece of engineering. They can produce a considerable amount of power, allowing an aircraft to fly, but still remain relatively light. But aircraft engines need a range of supporting systems to run smoothly, from removing the toxic exhaust gases to cooling to avoid overheating. These supporting systems will be explored in the next few chapters.

CHAPTER 5: **EXHAUST SYSTEM**

Imagine flying along on a cool winter's day. To combat the cool conditions, you have the heating system pumping warm air into the aircraft cabin, which is keeping you nice and snug. However, after flying for only a short period, you begin to experience a severe headache and decide to head back early to your home base. You may be surprised to learn the source of your headache is an invisible toxic gas that has snuck into the heated air through a faulty exhaust system.

Exhaust gases are the waste product that is produced when the fuel-air mixture is ignited. You may recall from the previous chapter that the exhaust gases are swiftly pushed out of each cylinder after combustion, but that's not the end of the journey for the exhaust gases. Once they exit through the exhaust valves of each cylinder, they enter the exhaust system, designed to safely dispose of the exhaust gases away from the aircraft cabin. Exhaust fumes contain several toxic gases, however, one in particular is of great concern – **carbon monoxide**, as shown in the following case study:

> In September 2016, a pilot was conducting a local flight in Alaska. A witness reported seeing the aircraft make a number of steep turns at low levels before plunging into the ground. The post-accident investigation of the aircraft's exhaust system revealed it had a large crack, which would have allowed exhaust gases to enter the cockpit. Toxicology tests showed the pilot had 48% carbon monoxide in his blood. The pilot was a non-smoker, which would typically result in less than 3% of carbon monoxide in the blood. The pilot's severe impairment from carbon monoxide poisoning likely resulted in the pilot losing control of the aircraft.

Carbon monoxide is a hazardous gas, but it can be very difficult to detect as it is colorless, tasteless and odorless. Carbon monoxide attaches itself to the body's red blood cells, severely limiting oxygen's ability to be transported around the body, even if plenty of oxygen is available (e.g. at sea level). Carbon monoxide is produced through incomplete combustion, with the exhaust gases of an aircraft engine being the most common source of carbon monoxide for pilots. Carbon monoxide poisoning is a form of hypoxia, as a result most of the symptoms of hypoxia can occur, such as impaired judgment, confusion, headache and difficulty performing basic tasks. One of the final symptoms of carbon monoxide poisoning is unconsciousness and then eventually death.

In most small aircraft the engine is located in front of the pilot, therefore, the exhaust system needs to safely divert the gases away from the aircraft cabin. When the exhaust gases exit each cylinder, they enter the **exhaust manifold** – which is basically a series of pipes that the exhaust gases travel in. The exhaust gases are then ducted down to the bottom and back of the engine, and expelled safely below the aircraft cabin, as shown in Figure 5.1. Any joins in the exhaust manifold should be checked carefully during a preflight inspection to ensure they are properly sealed.

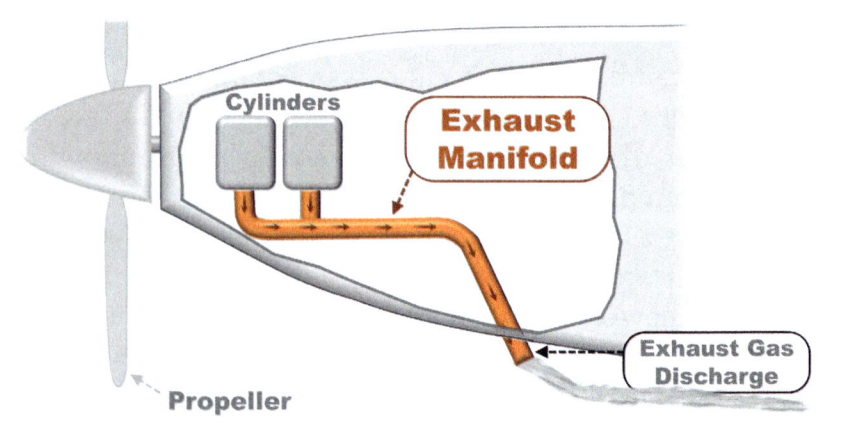

Figure 5.1: *Example of an exhaust system of a small aircraft.*

This all sounds simple, with little room for the exhaust gases to enter the aircraft cabin. However, the heating system of some aircraft may offer a route for exhaust gases – and therefore carbon monoxide – to enter the aircraft cabin. The exhaust manifold becomes very hot. Rather than let all this heat go to waste, some aircraft use it to heat air for the aircraft cabin. Fresh outside air is ducted around the outside of the warm exhaust system (inside a heat shroud or heat exchanger) and this heated air is then directed towards the cabin, as shown in Figure 5.2. Usually, the exhaust gases and the cabin air remain safely separated, however, if there is a crack in the exhaust system, toxic gases may enter the cabin via the heating system. If you smell exhaust fumes when the cabin heat is turned on, it should be treated as a warning sign of potential carbon monoxide in the air. You should:
- turn off the cabin heat
- increase the rate of fresh air by opening air vents
- use supplementary oxygen if it is available.

Importantly, carbon monoxide can remain in a person's blood for several days; therefore, even after landing the symptoms of carbon monoxide poisoning can persist. Due to the significant hazard of carbon monoxide, all aircraft are required to be fitted with a carbon monoxide detector to warn pilots of this hidden danger.

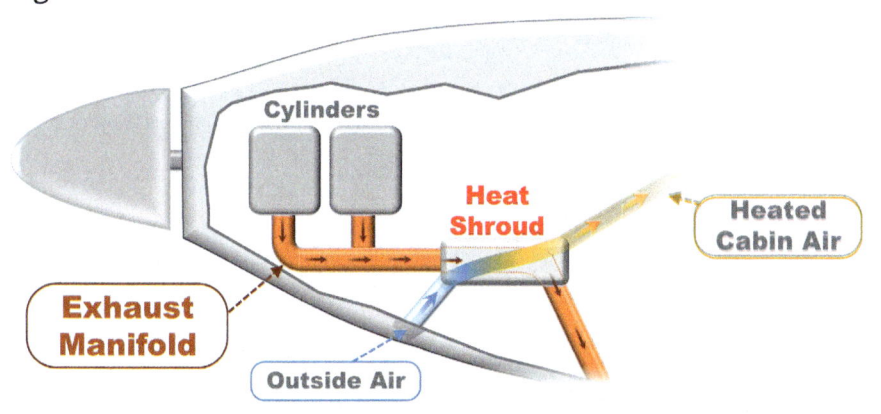

Figure 5.2: *Example of a heating system found in some small aircraft.*

The exhaust system in small aircraft is relatively simple, providing a safe passage to discharge the toxic gases away from the aircraft cabin after combustion. However, it is important pilots carefully inspect the exhaust system before a flight, as a faulty system can severely impact a pilot's ability to operate an aircraft safely.

CHAPTER 6: **INDUCTION SYSTEM**

In the previous chapter you explored the system to remove the waste product from the engine (exhaust system), whereas now you will look at the other end of the process, the induction system. The induction system delivers the fuel-air mixture to the cylinders for combustion. The keyword in the previous sentence is *mixture*. Fuel by itself will not burn without oxygen (air), therefore one of the key tasks of the induction system is to mix the fuel with air at the correct ratio. The ideal ratio is about 1-part fuel to 12-part air by weight (1:12). This ratio results in all the fuel and all the oxygen being used during combustion. Combustion in the cylinders can occur when the fuel is between about 1:9 (known as a rich mixture) and 1:18 (lean mixture). A rich mixture means there is excess fuel after combustion, therefore some unburned fuel will remain. A lean mixture means there is a shortage of fuel, therefore some oxygen remains after combustion. The induction system is designed to not only deliver the fuel-air mixture to the cylinders but also allow the pilot to adjust the ratio to suit the needs of the flight.

The two main types of induction systems found in small aircraft are the **carburetor** system and the **fuel injection** system. The starting point of both systems is collecting air from the outside environment. This is achieved using an **air intake**, typically found at the front of the aircraft, as shown in Figure 6.1. The air will first pass through an air filter to remove various foreign objects, such as dust. Just in case the air filter becomes blocked, an aircraft may also have an alternate air source, usually from inside the engine cowling (the engine cowling is the covering around the engine). The alternate air system may operate manually (e.g. a lever in the cockpit operated by the pilot) or automatically.

Figure 6.1: *Example of an air intake.*

The air intake is an important area to check during a preflight inspection. If it becomes blocked and an alternate source of air is unavailable, the engine may fail, as shown in the following case study:

> On the 20th of August 2019, a single-engine Aero Commander 200 took off on a maintenance check flight over Michigan. At around 300 feet the engine failed, leading to the aircraft crash landing a short distance from the aerodrome. The post-accident investigation found part of the air filter had not been properly installed, resulting in the filter becoming dislodged and blocking the air intake. With the air intake blocked and no alternate air supply available, insufficient air was mixed with the fuel, leading to the engine failure.

Carburetor

The next stage of the induction system requires the air to be mixed with fuel before being delivered to the cylinders. The carburetor is not normally used in modern small aircraft engines; however, many old aircraft remain flying with carburetors. The carburetor can suffer from several issues, which if not managed correctly can lead to an engine failure. The most common type of carburetor found in small aircraft is known as the **float-type carburetor**. As the name indicates it contains a float that helps control the amount of fuel being delivered to the engine, as shown in Figure 6.2.

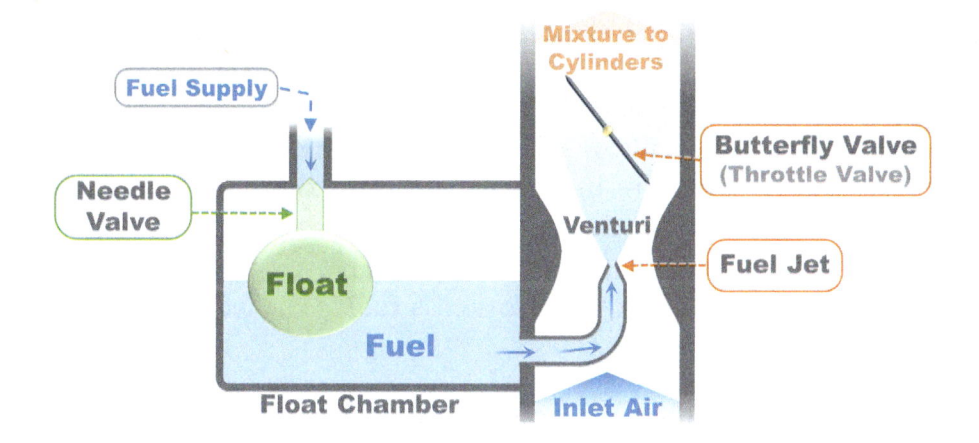

Figure 6.2: *Cross-section of a basic float-type carburetor.*

The air from the air intake first flows through a **venturi**. A venturi is a narrow section in the air passage that causes the air to accelerate (due to the air squeezing through a narrow section). Importantly, the venturi also results in the air pressure reducing in the narrowed section (following Bernoulli's principle of increased velocity = decrease static pressure). The amount of air flowing through the venturi (and towards the cylinders) is controlled by a **butterfly valve** (also called a throttle valve), which is located just upstream of the venturi.

In the venturi is a **fuel jet**, which is connected to the fuel in the **float chamber**. Fuel will naturally flow from areas of high pressure to low pressure (just like air will naturally flow out of a tire if a hole develops in it). With the fuel jet located in a low pressure area, it will force fuel to flow through it from the float chamber. As the fuel flows through the fuel jet, it mixes with the air, creating a fuel-air mixture that heads towards the cylinders. The fuel jet may also contain an atomizer and diffuser that helps vaporize the fuel, and therefore improve mixing with the air.

How quickly fuel flows through the fuel jet (and mixes with the air) will be determined by the position of the butterfly valve. The butterfly valve is directly connected to the throttle lever in the cockpit. At high power settings, the butterfly valve is wide open, allowing more air to flow through the venturi. This results in a lower drop in pressure in the venturi, and therefore more fuel being sucked through the fuel jet. When less power is required (e.g. during a descent), the butterfly valve is partially closed, reducing the flow of air, therefore less fuel is sucked through the fuel jet.

The float-type carburetor gets its name from the float that is located in the float chamber. The float chamber is where a small amount of fuel is held before being delivered to the fuel jet. The amount of fuel held in the chamber is maintained at a constant level with the help of the float. The float operates a needle, that either allows more fuel to enter the chamber (from the main fuel system) or stops fuel from entering the chamber. If the engine is using plenty of fuel (e.g. throttle fully open in a climb), the fuel in the chamber will drop, along with the float. This results in the needle valve opening, allowing more fuel to enter the chamber. When less fuel is being used, the fuel level and float will rise in the chamber. This will result in the needle closing the valve, therefore preventing more fuel from entering the chamber.

The carburetor system also includes several components to help maintain a smooth and constant fuel flow across a range of engine settings.

Accelerator Pump: When the throttle is opened quickly to maximum power, air will swiftly rush through the air passage, but the fuel in the float chamber is slower to react. Without an accelerator pump, a lag of power would be experienced due to insufficient fuel being mixed with the air. Not an ideal situation, especially when a pilot requires maximum power, such as during a go-around. To overcome this issue, a small manual pump is installed in the float chamber (accelerator pump), which is connected to the throttle linkage. The accelerator pump provides an extra squirt of fuel when the throttle is swiftly opened.

Idling System: When the engine is idling (low power setting, with the engine turning slowly), the pressure difference between the venturi and the float chamber is not enough to suck fuel through the main jet. To solve this issue, the carburetor has an idling system that delivers a small amount of fuel slightly upstream of the butterfly valve, as shown in Figure 6.3. The small passage allows fuel to bypass the venturi and deliver fuel to the other side of the butterfly valve (where the pressure is a little lower due to air passing through the cylinders). This provides enough fuel to keep the engine running when idling. When sufficient air is passing through the venturi (e.g. throttle increased for more power), fuel stops running through the idling passage.

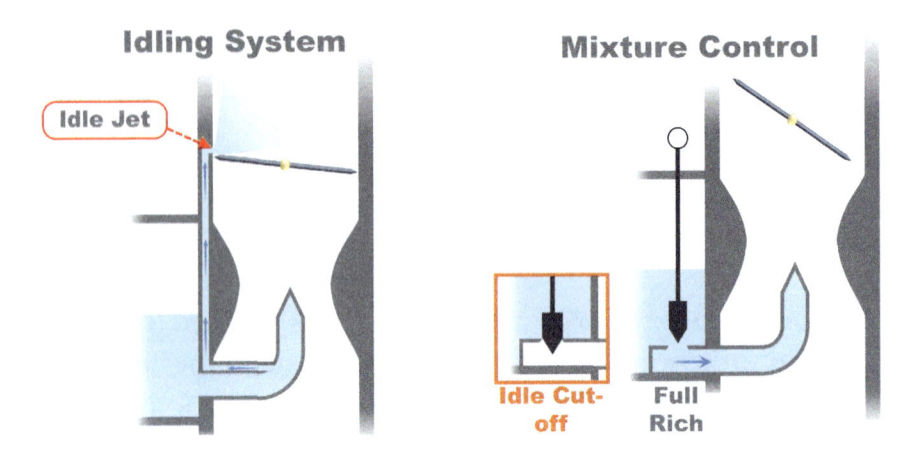

Figure 6.3: *Idling system and mixture control in a simple float-type carburetor.*

Mixture Control: The carburetor is generally calibrated for sea level conditions, or more specifically sea level air density. This means the ratio of air to fuel is usually correct at sea level when the mixture control is in the full rich position (the mixture control is typically the **red** control next to the throttle, as shown in Figure 6.4). But moving higher in the atmosphere results in a lower air density, meaning there is less air being mixed with the same amount of fuel that enters each cylinder. An over-rich mixture can result in the engine running rough and losing power due to spark plug fouling (carbon build-up on the spark plug due to lower temperatures during combustion). To ensure the correct fuel-air mixture is used, the mixture control can be adjusted to lean the fuel. The mixture control moves a small needle in the float chamber that restricts the amount of fuel to the fuel jet, as shown in Figure 6.3. It is also important to avoid an over-lean mixture (not enough fuel for the air available). Over-leaning can result in high cylinder temperatures, leading to detonation or pre-ignition. Over-leaning may occur during a descent from a high altitude, where the engine was leaned, but a richer mixture setting is required now the aircraft is entering an area with higher air density.

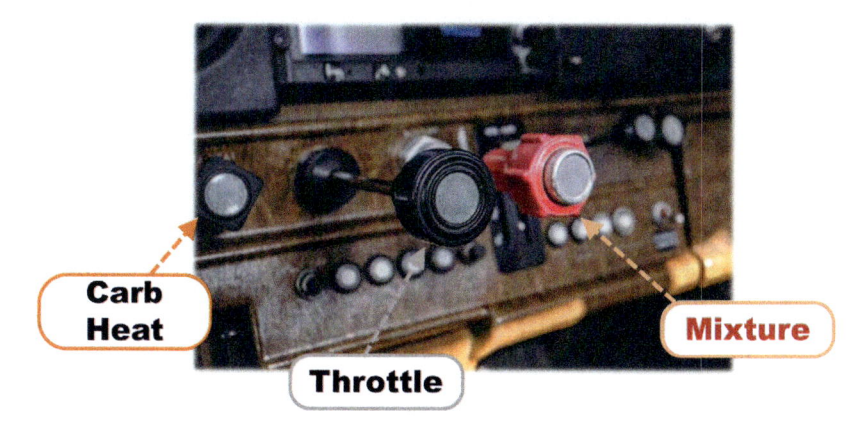

Figure 6.4: *Basic engine controls (fixed pitch propeller).*

When operating below 2,000 feet, during take-off, landing and descending, the mixture control will usually be kept in the full rich position (mixture control fully forward). A full rich mixture is also recommended when cruising with a high-power setting (more than 75%), as the rich mixture is required to help cool the cylinders (normal cruise power is about 55 to 65%). Leaning is recommended once you have settled into your cruising altitude, especially above 5,000 feet, which can greatly improve fuel efficiency. To lean the mixture, slowly move the mixture control towards the lean position. You should notice the engine revolutions per minute (rpm) slowly increase due to the correct fuel-air mixture being regained. A point will be reached where the rpm starts to reduce, and the engine may start to run rough. This means you have gone too lean (not enough fuel in the mixture); therefore, you should move the mixture control towards the rich position a little. The engine should be run on the rich side of the optimum position to avoid overheating. Some aircraft may have an exhaust gas temperature (EGT) gauge which measures the temperature of the exhaust gases. This can also be used when leaning, with the optimum mixture position obtained at peak EGT. The leaning procedure should be repeated every time there is a change in cruising altitude or power setting.

Idle Cut-off: The mixture control is also used to shut the engine down – sometimes called an idle cut-off. When the mixture control is closed, the needle stops fuel from leaving the float chamber, with the engine stopping due to a lack of fuel, as shown in Figure 6.3.

The float-type carburetor suffers from two main problems. First, it does not function well during abrupt maneuvers, such as aerobatics. The flow of fuel running into the float system largely relies on gravity, therefore during abrupt maneuvers the flow of fuel can be interrupted. Second, the area around the venturi can become blocked with ice. **Carburetor icing** is sometimes called fuel or throttle ice. The tricky aspect of carburetor icing is it can occur on a clear day when the outside air temperature is well above freezing (0°C / 32°F). The following case study highlights the dangers of carburetor icing:

> On the 9th of October 2003, a single-engine Cessna 172 took off on a sightseeing flight over Toronto, Canada, with a pilot and three passengers on board. Shortly after leveling off at 2,000 feet, the engine began to lose power. The pilot conducted a range of checks, including briefly selecting carburetor heat, after which the pilot observed a further decrease in power and therefore turned the carburetor heat off again. The engine was not producing sufficient power to maintain level flight, therefore the pilot conducted a successful forced landing. With a temperature of 23°C (73°F) and a high moisture content in the air, carburetor icing was the most likely reason for the loss of power. Unfortunately, the pilot did not leave the carburetor heat on long enough for it to be effective.

As air enters the venturi the temperature drops, and this drop can be quite considerable, sometimes up to 35°C (63°F). If the temperature in the carburetor drops below freezing, ice can form, as shown in Figure 6.5. There is also a high risk of ice formation around the butterfly valve during low power settings, such as a descent. During low power settings, the butterfly valve is partially closed creating a further venturi effect and therefore additional cooling.

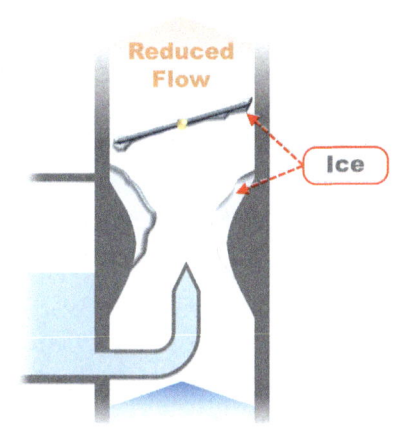

Figure 6.5: *Carburetor icing can form around the venturi and butterfly valve.*

Carburetor ice is more likely to form when the outside air temperature is about 5°C to 25°C (41°F to 77°F) but can occur in warmer temperatures when a high amount of moisture is in the air (high relative humidity). The first sign of carburetor icing is a rough running engine and degraded power. If left untreated, the engine will stop due to fuel starvation. Prevention is the key to avoiding carburetor icing in the first place. When operating for extended periods with a low power setting, apply full power occasionally and turn on carburetor heat (carb heat), which was shown in Figure 6.4. Carb heat will result in warm air entering the carburetor, melting any ice. Carb heat will also reduce engine power, due to the warm, less dense air being used for combustion, therefore it is advisable not to operate the whole flight with carb heat on. Carburetor ice can also form on the ground, especially when taxiing with a low power setting when the engine is still cold (e.g. first flight of the day). A proper engine run-up should help prevent ice build-up prior to take-off.

Fuel Injection

Many of the issues of a carburetor have largely been eliminated with the introduction of the fuel injection system – found in most modern small aircraft. The key aspects of the fuel injection system include:

- the fuel is delivered under pressure (using a fuel pump)
- the fuel is mixed with the air by injecting it into the inlet manifold just before each cylinder.

The initial components of a fuel injection system are similar to a carburetor system, whereby air enters an air intake and passes a butterfly valve, however, the fuel is not mixed with the air at this stage but continues towards each cylinder. A key component of a fuel injection system is a **fuel control unit (FCU)**, which is linked to the throttle and mixture controls in the cockpit. The fuel control unit regulates how much fuel should be delivered to each cylinder, depending on the throttle and mixture settings selected by the pilot. From here fuel travels to a **fuel manifold unit (FMU)**, where fuel is distributed to separate fuel lines that deliver fuel to each fuel discharge nozzle. The nozzles are usually located just before the inlet valve of each cylinder, as shown in Figure 6.6. It is at this point the fuel is mixed with the air and enters each cylinder (some systems may inject fuel directly into the cylinder, known as a direct fuel injection system). A fuel injection system will usually have two types of fuel pumps to help deliver the fuel:

- An **engine-driven** fuel pump that operates when the engine is running (and is powered by the engine).
- An **electric** fuel pump (also called a boost or auxiliary pump) is used when starting the engine and as a backup if the engine-driven pump fails.

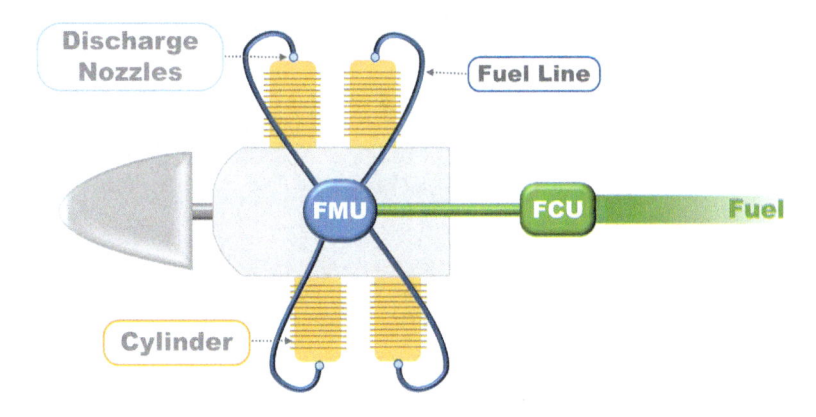

Figure 6.6: *Basic components of a simple fuel injection system, including the fuel control unit (FCU) and fuel manifold unit (FMU).*

As the fuel is not injected until after the venturi, the risk of icing is eliminated, and the system will continue to operate during abrupt maneuvers. A fuel injection engine also tends to be more efficient as each cylinder can be provided with the correct fuel-air mixture ratio, unlike a carburetor system that struggles to deliver an even mixture to all cylinders. The two main issues with a fuel injection system are hot starting and fuel contamination:

- Vapor locking in the fuel lines can make it difficult to start a hot engine. This will be discussed in more detail in chapter 10 (*engine management*).
- The fuel injection system is more susceptible to contamination in the fuel, such as dirt or water, that can block the fine fuel lines or injector nozzles.

Delivering the correct fuel-air mixture into each cylinder is essential for smooth engine operations. It is important pilots are aware of the type of induction system in their aircraft and its limitations. This is especially the case for an aircraft with a carburetor, where carburetor icing is a constant threat, even on a relatively warm day.

CHAPTER 7: **IGNITION SYSTEM**

In a spark ignition engine, with the correct fuel-air mixture being delivered to the cylinder, a spark is now required to ignite the mixture. The spark is provided by the ignition system, whose goal is to deliver an electrical current to each spark plug, at the correct time. In small aircraft the ignition system usually contains a magneto, high tension leads, spark plugs and an ignition switch – collectively called a magneto ignition. The magneto ignition is relatively old technology, but as it is independent of the aircraft's electrical system, it will continue to run even if the aircraft's electrical system fails. Most cars have an electrical ignition system, which is slowly making its way into aviation but is still rare in small aircraft (this will be covered in more detail in chapter 10, with the FADEC system). Aircraft with a compression ignition engine (e.g. diesel engine) do not require an ignition system, as they do not have spark plugs.

Magneto Ignition

The first part of this system is called a **magneto.** The magneto is a self-contained unit driven by the engine and generates a high voltage electrical current with a rotating magnet. This electrical current is then delivered to each spark plug by **high tension leads**. These leads are designed to ensure the electrical current does not escape to other parts of the aircraft, which could be very dangerous. At the **spark plug** inside the cylinder head, the electrical current jumps across two electrodes, causing a spark, thereby igniting the fuel-air mixture. As the magneto is driven by the engine, it operates whenever the engine is running and will continue to provide a spark even if the aircraft's electrical system fails. But what if a fault occurs within the magneto system, will the engine stop? Most aircraft will have a **dual ignition system** – two completely independent systems. Each cylinder will have two spark plugs, each being fed by a different magneto system (each with its own magneto and high tension leads), as shown in Figure 7.1. Dual spark plugs in each cylinder also improves the combustion of the fuel-air mixture, allowing more power to be delivered. If one ignition system fails, the other one will keep the engine running, although with a slight power reduction.

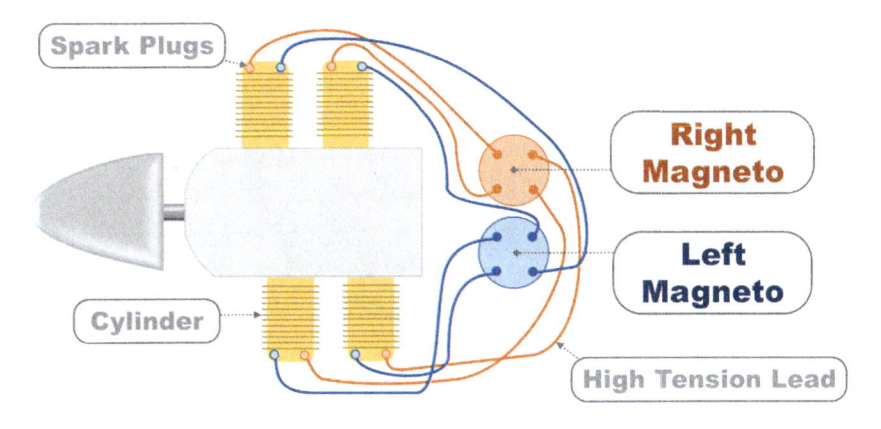

Figure 7.1: *Basic components of an ignition system.*

The ignition system is also designed to ensure the spark occurs at the correct time (usually just before the piston reaches the top-dead-center position prior to the power stroke). The following case study highlights the consequences of incorrect spark timing:

> On the 8th of August 2008, a single-engine Cessna 207 was conducting a flight in Canada after scheduled engine maintenance. Shortly after take-off, the temperature in the cylinders rose rapidly, well beyond the normal range, and the engine began to vibrate. The pilot turned back towards the runway but was unable to maintain altitude and was forced to make a landing on a nearby road. The engine issue was traced to the incorrect timing of both magnetos, which had just been refitted into the aircraft. The magnetos should

be timed to ignite the fuel-air mixture prior to the power stroke at 22 degrees before top-dead-center. However, when the magnetos were refitted, they were accidentally timed to 50 to 60 degrees before top-dead-center. Such advanced timing would have led to pre-ignition and detonation in the cylinders, resulting in high cylinder temperatures and power loss.

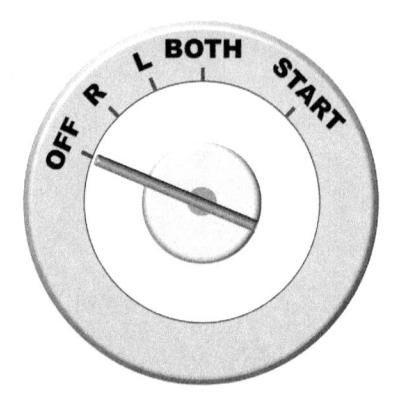

Figure 7.2: *Example of an ignition switch that may be found on a small aircraft.*

The magneto system is usually operated by an ignition switch in the cockpit, as shown in Figure 7.2. The switch has five points:
- OFF
- R (right)
- L (left)
- BOTH
- START

When a pilot selects either LEFT or RIGHT on the ignition switch, only one magneto will operate (i.e. if LEFT is selected, only the left magneto operates), whereas when BOTH is selected, both magnetos are operating, and this is the normal ignition switch position during a flight. A common check before take-off is to make sure each magneto system is operating. When switching from BOTH to RIGHT, and from BOTH to LEFT, a small decrease in engine revolutions per minute (rpm) should be expected. The allowed reduction will be stated in the Pilot's Operating Handbook. If the engine stops when switched to one magneto, or a large rpm drop is experienced, a fault has occurred in the system and the aircraft should not be flown. You should also be cautious of no rpm drop at all, as this indicates the magneto is still 'live', even when it is turned off. For example, when switching to RIGHT, if no drop in rpm is experienced, then the left magneto may still be running when it shouldn't be. This can be very dangerous if someone moves the propeller when the engine is turned off (e.g. moving the aircraft into a hanger). With the magneto still 'live' in the off position, the engine may inadvertently start.

Starting

Once the engine is running, the ignition system is largely self-driven, that is, the engine turns the magnetos which creates the electrical current for the spark plugs. But when a pilot is trying to start the engine, the magneto ignition system needs a little help. This is achieved with a starter system, that starts turning the engine using electrical power and an impulse coupling to create the spark:

> Most small aircraft have a direct-cranking **electric starter system**. Basically, this is an electrical motor that turns the engine when the pilot selects START on the ignition switch. The starter will automatically disengage when either the engine is running at normal speeds or the START switch is released on the ignition switch. Electrical power is supplied from the battery (or external power unit, if the aircraft is plugged in). When the starter is engaged, the engine will turn over relatively slowly (about 120 rpm, compared to 800 - 1,000 rpm at normal idle engine speeds).

> An **impulse coupling** is used to provide enough energy to deliver a spark, even when the engine is turning over very slowly when the starter is engaged. Just as important, the

impulse coupling also helps deliver the spark at the right time. The impulse coupling is a very clever and simple device that is designed to operate when the engine is trying to start, but not operate at normal engine speeds (as the normal magneto system will operate). At low engine speeds, the impulse coupling holds back (retards) the magnet in the magneto, and at the same time it winds up an internal spring in the impulse coupling. Just before the piston in the cylinder begins the power stroke, the spring releases the magnet, allowing the magneto to generate sufficient electrical current for the spark plugs.

The ignition system found in small aircraft has not changed much through the years. This is a testament to the reliability of this key component that helps keep the engine running throughout a flight. But like other components in the aircraft, pilots need to remain alert for any signs of malfunctions, and never operate an aircraft with a faulty ignition system.

CHAPTER 8: **OIL AND COOLING SYSTEMS**

An aircraft engine is a heat machine. When the spark plugs ignite the fuel-air mixture, the temperature inside each cylinder becomes incredibly hot. Some of this heat is expelled through the exhaust system, but the rest needs to be removed to prevent the engine from overheating, which could result in the engine failing. It is not just the combustion process that creates heat, but also friction from the moving parts within the engine. To keep the engine operating smoothly, aircraft have lubrication (oil) and cooling systems. These systems not only remove heat from various parts of the engine but also help prevent the fast-moving parts from rubbing together.

Oil

The key component in the lubrication system is oil. You will see shortly that oil constantly flows through the engine to ensure it remains lubricated and takes away unwanted heat. The oil also carries away foreign material, such as dirt, and provides a seal between the cylinder walls and the piston, therefore preventing the fuel-air mixture from escaping into the crankcase. The aircraft engine is very demanding on the oil due to the range of temperatures at which it must operate. To be effective, the oil needs to maintain a certain level of viscosity. **Viscosity** means how easy the oil will flow at a specific temperature. Low viscosity oil will flow more easily, whereas high viscosity oil will flow slower. What is the ideal oil viscosity in the aircraft engine? Each engine will have a specific recommended oil type, providing a balance between not too thick – which would make it difficult to flow through the engine – and not too thin – which reduces the effectiveness of the oil to lubricate. The challenge is viscosity changes with temperature:

- High temperatures make the oil less viscous (flows quicker).
- Low temperatures make the oil more viscous (flow slower).

Engine temperatures can change considerably, therefore the oil in the engine needs to operate over a wide range of temperatures. To help ensure the engine uses the correct oil type, oil is graded under a **Society of Automotive Engineers (SAE)** rating system. For the same engine, the recommended oil type may vary. For example, an aircraft that operates in cold climates may require an oil of lower viscosity compared to an aircraft that operates in warmer climates. Pilots need to ensure the aircraft has the correct amount of oil and the correct type.

Lubrication System

With the correct oil in the aircraft, let's explore the engine's lubrication system. There are two main types of systems found in small aircraft: a wet sump and dry sump system. The main components are similar, with the key difference being where the oil is stored when it is not circulating around the engine. In both systems, oil is first distributed to various parts of the engine, as shown in Figure 8.1. The oil then ends up at the bottom of the engine (around the bottom of the crankcase), an area also known as the sump.

- In a **wet sump** system, this is where the oil remains until it is pumped back to the top of the engine to be used again.
- In the **dry sump system**, when the oil reaches the bottom of the engine, a scavenger pump pumps the oil into a separate storage tank, leaving the bottom of the engine 'dry'.

Most small aircraft have a wet sump system, whereas a dry sump system is more likely to be found in aerobatic aircraft that operate in unusual attitudes (e.g. upside down).

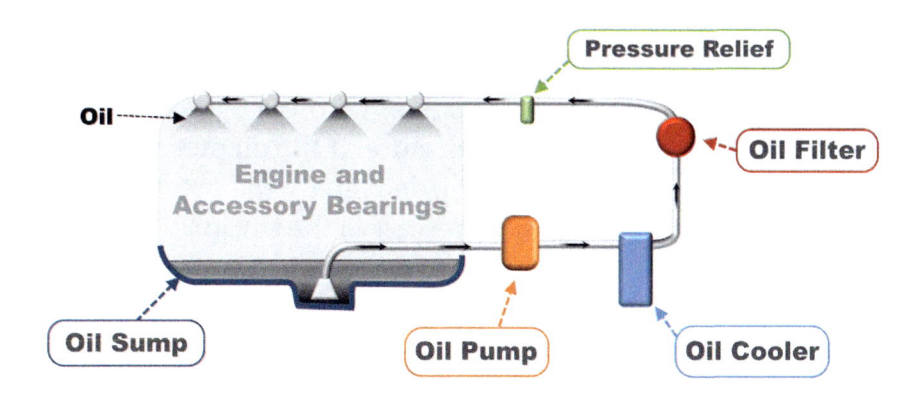

Figure 8.1: *The typical components of a wet sump system.*

Each system will contain an engine-driven oil pump that pumps the oil from its storage location, then routes it through the engine. Along the way, the oil will also pass through an oil filter, oil cooler and pressure relief valve.

> *Oil Cooler:* The oil absorbs a lot of heat from the engine. Some of this heat will be lost when the oil is sitting in the sump, but usually, further cooling is required using an oil cooler before the oil passes through the engine again. If the oil is already cool it will bypass the cooler, whereas if it is warm, it is directed through the oil cooler.
>
> *Oil Filter and Screen:* You may recall, one of the functions of the oil is to collect various contaminants in the engine (e.g. dirt and other foreign material). This is why engine oil is usually golden when new but turns black the longer it remains in the engine. The contaminants need to be collected from the oil before going through the engine again. An oil system will contain several **screens,** for example a

screen is usually fitted to the point oil is pumped out of the sump. Screens have a fine mesh that pick up larger objects in the oil. The **oil filter** is then used to collect finer material. The oil filter also contains a bypass valve that allows oil to bypass the oil filter if it becomes clogged. Contaminated oil running through the engine is preferred to no oil at all.

Pressure Relief Valve: The oil passes through a pressure relief valve that regulates the oil pressure. If the oil pressure is too high (e.g. too much oil is being delivered), the pressure relief valve will allow excess oil to return to the sump.

Oil Pressure and Temperature

Pilots can determine the health of the lubrication system, and therefore the engine with oil pressure and oil temperature gauges, as shown in Figure 8.2.

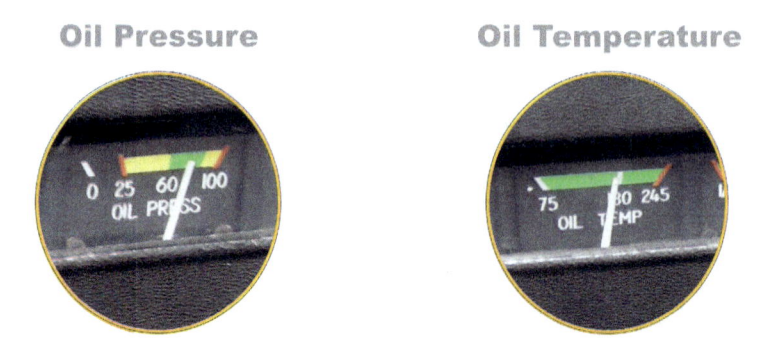

Figure 8.2: *Examples of oil pressure and oil temperature gauges.*

The oil is delivered under pressure to the engine, which is usually displayed in pounds per square inch (psi). Green indicates normal operating range and red for minimum and maximum pressures, as shown in Figure 8.2. Oil temperature is also displayed on a gauge, again with green and red markings. Oil pressure changes relatively quickly, for example after the engine has started oil pressure should sit within the green range relatively quickly. However, oil temperature changes slowly, gradually rising after the engine starts (especially if the engine is cold). Abnormal readings in these two gauges should be treated seriously:

Low Oil Pressure: This is a sign the engine is not well. There may be a blocked oil line or a broken seal. Whatever the cause, there is not enough oil circulating in the engine. You should divert to the nearest aerodrome, keep power settings to a minimum and be prepared for the engine to fail. Low oil pressure will normally be associated with a rising oil temperature, and a very high oil temperature with a low oil pressure indicates the engine may fail at any moment.

High Oil Pressure: This may be due to the pressure-relief valve not working. This puts strain on the oil system, possibly leading to a rupture at a weak link (e.g. an oil line), resulting in the oil being pumped out of the engine. You should try operating the engine at a lower power setting and land at the nearest suitable aerodrome.

Fluctuating Oil Pressure: Oil pressure may fluctuate during a flight. Provided it remains within the green range it should not be too much of a concern. This may be due to a faulty bypass valve or pressure relief valve. If it is fluctuating out-of-limits (e.g. in and out of high oil pressure), you should treat it the same as the out-of-limits action.

High Oil Temperature: If pressure is still normal, this one can be difficult to determine the cause. This may be due to the day being very hot or flying with a high-power setting (e.g. long climb). Reduce power and place the aircraft in a normal level attitude to see if the temperature returns to normal. If not, a more serious problem may be occurring, such as too little oil circulating in the system, and you should divert to a suitable aerodrome.

Pilots should check oil levels during their normal preflight inspection. The following case study highlights the consequences of operating with insufficient oil:

> On the 9th of June 2018, a pilot was conducting a cross-country flight across the United Kingdom, in a single-engine Cirrus SR22. About eight minutes into the flight, the engine began running rough, followed by a loss of power and grey smoke coming from the engine. The pilot was unable to maintain altitude, therefore operated the Cirrus Aircraft Parachute System at 800 feet and safely descended into a field below. The aircraft likely had insufficient engine oil, resulting in the temperature at the bottom of the engine becoming so hot that parts of the engine started to melt.

Cooling Systems

The oil and exhaust systems do a great job at collecting and removing heat from the engine. But more cooling is still required, especially around external engine surfaces, such as the outside of the cylinders that become very hot. If the engine becomes too hot, it can lead to a loss of power, excessive oil consumption, detonation in the cylinders and engine damage. In most small aircraft, extra cooling is provided by directing outside air around these hot components, known as **air-cooling**. On an average day, the outside air temperature where aircraft fly is pretty cold, for example, if you popped outside the aircraft at 10,000 feet, you would find the temperature to be about -5°C (23°F). Cool outside air flows into the engine cowling through the **air inlet** located behind the propeller hub, as shown in Figure 8.3. Once inside the engine cowling, the air is directed to the hottest parts with the help of **baffles.** This hot air then exits the aircraft through an opening in the cowling found at the back and lower part of the engine.

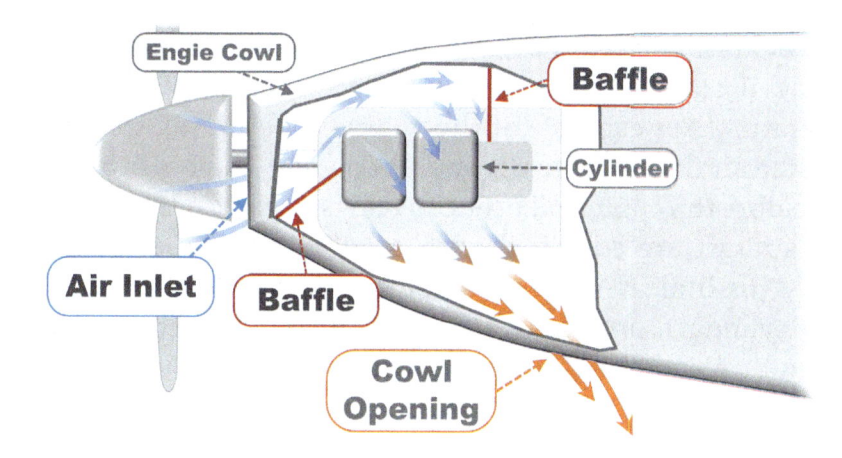

Figure 8.3: *Example of an air-cooling system.*

Some engine parts have design features to help with air-cooling, such as fins found on the outside of the cylinders, as shown in Figure 8.4. These features increase the area exposed to the airflow, therefore improving the cooling characteristics of the engine.

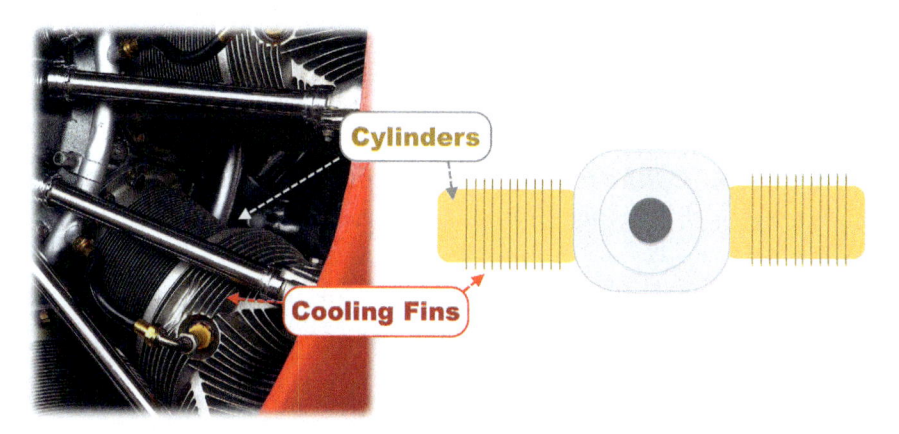

Figure 8.4: *Cooling fins on some engine parts, such as the cylinders, help with cooling.*

This type of cooling system is less effective in phases of flight with high power settings and low airspeeds, such as take-off, go-around and long climbs. Pilots can monitor the engine temperature through several gauges. The oil temperature gauge will provide a delayed indication of the engine temperature, but in some aircraft, this is the only means of assessing the engine temperature. For example, if the engine is becoming too hot due to limited air flowing into the engine cowling, there will be a short delay before the engine oil heats up and is displayed in the cockpit. Some aircraft are equipped with a **cylinder head temperature** gauge. This indicates the temperature at the cylinder head (top of the cylinder), providing a more immediate indication of the temperature of the engine, as shown in Figure 8.5.

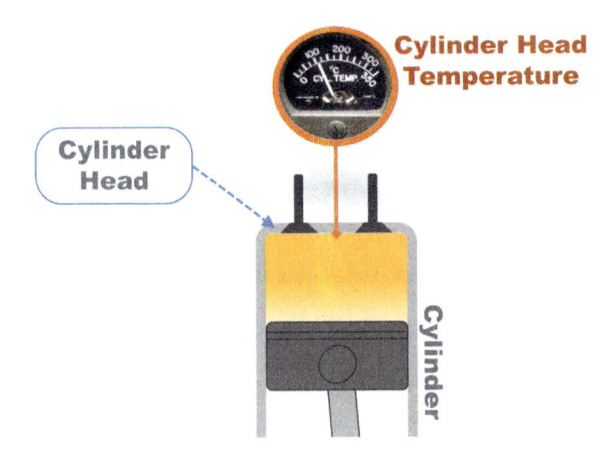

Figure 8.5: *Cylinder head temperature provides an accurate indication of the temperature of the engine.*

If the engine does start to overheat, pilots have a range of options available to bring the temperature back to the normal range:

Reduce power and *increase airspeed* (e.g. stop climbing or descending). These two options may not be suitable if the aircraft is flying close to terrain.

Enrich the fuel-air mixture (i.e. place the mixture control into the FULL rich position if it is not already). The extra fuel entering the cylinders (compared to when it is leaned), will help with cooling.

Open cowl flaps. Some aircraft are fitted with **cowl flaps** that are operated in the cockpit by a lever or switch. Cowl flaps are a hinged cover that opens and closes and are generally located towards the back and lower part of the engine cowling, as shown in Figure 8.6. When opened, it allows more hot air to escape the engine cowling (compared to the fixed opening). Most small aircraft do not have cowl flaps as the engine size and fixed cowling design allow for effective cooling. Cowl flaps are more likely to be fitted to aircraft with larger engines as they produce more heat and may have their cowling tightly fitted around the engine to reduce drag at high speed. Cowl flaps are usually opened during the take-off and climb phases to improve engine cooling and closed during cruise and descent.

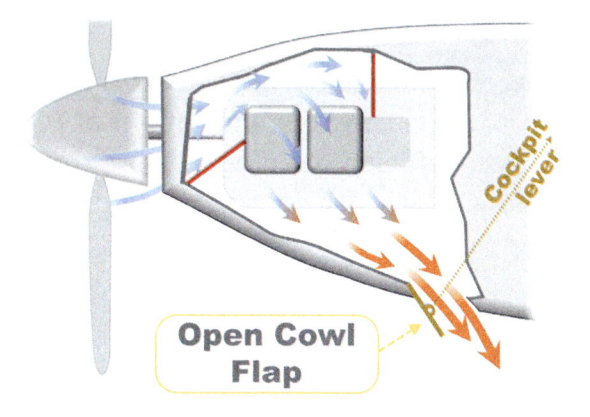

Figure 8.6: *Open cowl flaps help engine cooling.*

Although engine overheating is the main concern, pilots should also be on alert for overcooling. **Overcooling** can occur when the engine is hot and is suddenly cooled, sometimes called shock cooling. The abrupt temperature change can cause engine damage and is more likely in a high-speed descent (low power, high airflow into the engine). This is also another reason why cowl flaps are closed during a descent (in aircraft fitted with cowl flaps).

Liquid Cooling

Air cooling is the most practical way to cool the engine in small aircraft, providing a relatively simple, lightweight and reliable cooling method. However, the main disadvantage of air cooling is it is difficult to provide even cooling to all the cylinders, no matter how well the baffles direct the air. An increasing number of engines are turning to liquid cooling, which is common in cars. Liquid cooling involves liquid (coolant liquid) circulating around the outside of the cylinders, taking away excess heat, which is removed from the liquid through a heat exchange or radiator. This type of design is better at avoiding large variations in engine temperature, especially when airspeed is low (e.g. on the ground). However, due to more parts required, it is more complex and heavier, which is why it is not normally found in smaller aircraft (along with the fact that there is plenty of air usually available to provide air-cooling).

Aircraft engines become very hot, from the massive temperatures inside the cylinders to the constant movement of various parts. But with a reliable and efficient cooling system, this heat can be safely removed, allowing smooth engine operations.

CHAPTER 9: **PROPELLERS**

After the aircraft wing, the propeller (and the engine it is attached to) is one of the most important components of the aircraft. The propeller converts the power of the engine into a useful forward force known as thrust. Like the aircraft wing, the propeller is most efficient under certain conditions. You will see in this chapter there are some clever design features to help the propeller remain efficient over a wide range of flight conditions.

Basic Components

Let's start with the basic components of the propeller. Piston aircraft will have two or more propeller blades, with each essentially a rotating wing, as shown in Figure 9.1.

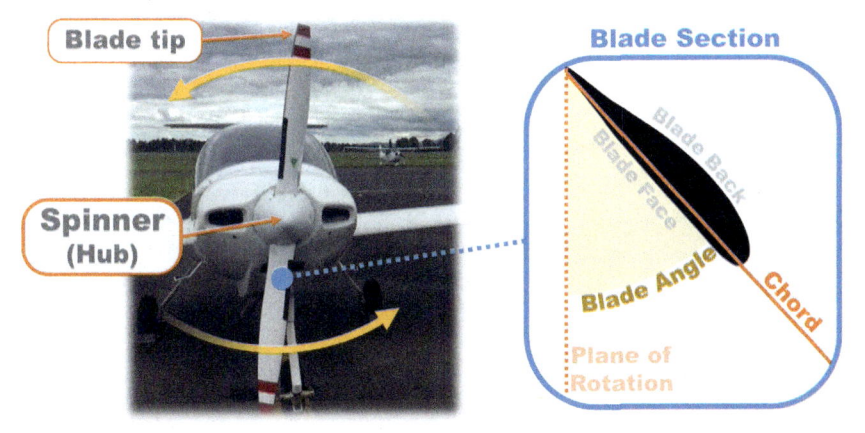

Figure 9.1: *Basic components of the propeller.*

Consider one propeller blade section at some distance from the spinner (hub), as shown in Figure 9.1. The blade section is an aerofoil (just like the aircraft wing), with one side of the blade fairly flat, whereas the other side is curved. The flat side is located on the side the pilot would see when sitting in the cockpit, which is also called the 'blade face', whereas the curved side is called the 'blade back'. The propeller blade also has a **chord line,** a line between the front and back of the blade section (known as the leading edge and trailing edge). **Blade angle** is the angle between the chord line and the plane of rotation, measured in degrees. The **plane of rotation** is an imaginary line that the propeller follows as it spins.

The propeller produces thrust when it is subjected to a **relative airflow (RAF).** The RAF is the airflow that acts parallel with and directly opposite to the propeller's path. The angle between the chord line and the RAF is known as the **angle of attack (AoA),** as shown in Figure 9.2. Angle of attack is extremely important and has a considerable impact on a propeller's ability to generate thrust, and at very high angles of attack, this thrust can suddenly collapse.

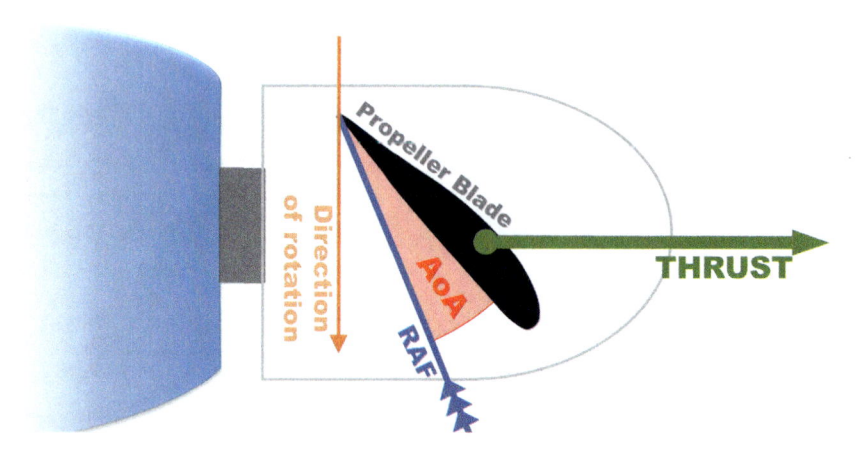

Figure 9.2: *Relative airflow (RAF) and angle of attack (AoA).*

Each propeller will be most efficient at a specific angle of attack (normally about 2-4 degrees). However, determining the RAF affecting the propeller is a little tricky, as it usually contains two components. Consider an aircraft that is stationary on the ground with the propeller spinning. With no forward speed, the RAF would simply be coming from the opposite direction to the plane of rotation, therefore the angle of attack would be the same as the blade angle. However, once the aircraft has some forward motion, the RAF of the propeller will have two components: **rotational velocity** and **forward speed** of the aircraft. Rotational velocity means how quickly a section of the blade is traveling. It is determined by how fast the propeller is spinning (revolutions per minute – rpm) and the specific location on the blade. The tips of the propeller have a faster rotational velocity compared to blade sections near the spinner, as shown in Figure 9.3.

Figure 9.3: *The rotational velocity of a section of the blade depends on the distance from the spinner (hub) and revolutions per minute (rpm).*

The forward component of the RAF is simply how quickly the aircraft is traveling. As a result, RAF and angle of attack will vary with changing airspeeds and rotational velocities along the propeller. Figure 9.4 shows the two components affecting a propeller section. The rotational velocity (RV) component is shown down the left side of each diagram, and the forward airspeed component is shown along the bottom. Once combined, RAF can be found.

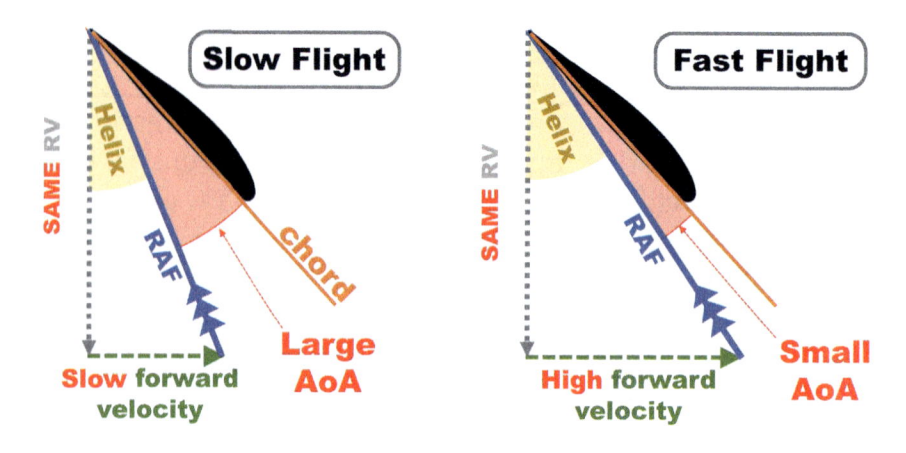

Figure 9.4: *For the same propeller rotational velocity (RV), angle of attack (AoA) will vary with changing forwards speeds of the aircraft. The angle between the RAF and rotational velocity is known as the helix angle*

For the same rotational velocity, as the aircraft travels faster, the RAF airspeed component becomes larger, and therefore the angle of attack is smaller, as shown in Figure 9.4. Likewise, at slower airspeeds, the angle of attack is higher. When the aircraft has some forward speed, a new angle is formed between the RAF and the plane of rotation, known as the **helix angle.** You could also say the angle of attack plus the helix angle equals the blade angle.

For the same forward speed, angle of attack will also change along different parts of the same propeller blade (assuming the blade angle remains the same). Consider Figure 9.5, which shows the RAF for the same propeller, traveling at the same forward speed, but at different points along the blade.

- The section near the middle of the blade has a lower rotational velocity, therefore the angle of attack is lower.
- The section near the tip has a larger rotational velocity, therefore the angle of attack is higher.

You will see shortly the blade angle varies along the length of the propeller to maintain a constant angle of attack.

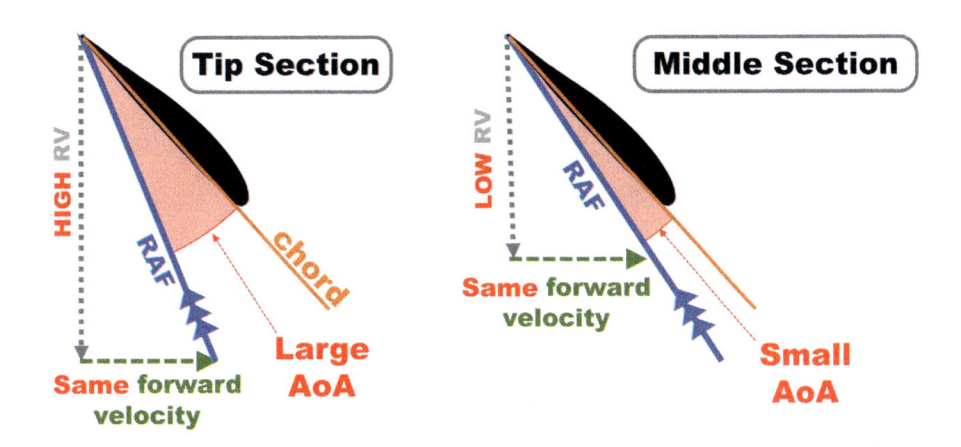

Figure 9.5: *For the same forward airspeed, the angle of attack (AoA) will vary along the length of the propeller due to changes in rotational velocity (RV), if blade angle remains the same.*

Blade Twist

The propeller is usually most efficient at an angle of attack of around 4 degrees. However, as the propeller spins faster the further it is from the spinner, the RAF constantly changes along the length of the propeller (due to the different rotational velocities at each propeller section). If the same blade angle was used for the whole propeller, the angle of attack would vary widely, as was shown in Figure 9.5. This would result in variable thrust being generated along the propeller. To overcome this issue, the propeller is twisted, meaning the blade angle changes along the length of the propeller, allowing a relatively constant angle of attack, therefore creating an even amount of thrust along the propeller. The twist angle is largest near the spinner, where the rotational velocity is the lowest, and smallest at the tip, as shown in Figure 9.6.

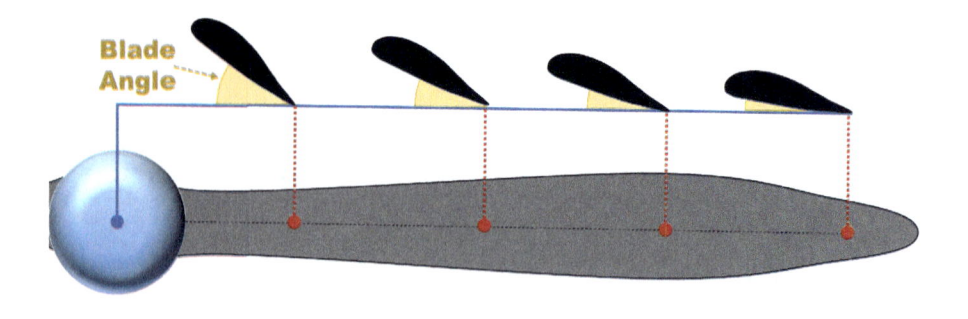

Figure 9.6: *The propeller blade is twisted to help produce an even amount of thrust along the blade.*

Most small aircraft are fitted with a **fixed pitch propeller**, meaning the blade angle cannot be adjusted during a flight. Despite the fixed pitch propeller being twisted, it will only be efficient at a given airspeed and rpm. Aircraft are normally fitted with either a climb or cruise fixed pitch propeller, depending on the type of flying the aircraft is intended for. Although a fixed pitch propeller is not efficient across all flight conditions, they are widely used due to their reduced complexity, meaning they are simpler and cheaper.

Propeller Efficiency

Angle of attack is not the only factor that will influence a fixed pitch propeller's ability to convert engine power into useful thrust. You will explore the constant speed propeller shortly, which can overcome the angle of attack issue. But other propeller design features can help, as shown in Figure 9.7, such as propeller diameter (the length of the propeller). A longer diameter means there is more propeller to generate thrust (just like a larger wing will generate more lift), and practically propellers are designed as long as possible. However, there are several limiting factors:

- The larger the diameter, the more engine power is required to turn the propeller. This means the engine power limits will be reached at some point (i.e. if the propeller is too long, the engine will not be able to turn it).
- Propeller diameter is often limited by ground clearance (distance between the propeller tip and the ground). A propeller striking the ground on take-off or landing is not ideal.
- If the diameter is too long, the tip speed can reach the speed of sound resulting in excessive noise, vibrations and reduced thrust.

Another design feature to improve propeller efficiency is to increase the number of blades or increase the width (chord) of the propeller blade. However, as discussed above, this would also require a more powerful engine to turn the propeller.

Figure 9.7: *Propeller efficiency can be improved by increasing the number of blades and having the propeller as long as practicable.*

Constant Speed Propeller

To overcome some of the efficiency issues of a fixed pitch propeller, some aircraft are fitted with a constant speed propeller. A fixed pitch propeller is only efficient at a specific airspeed and rpm. But the loss in efficiency can significantly hinder an aircraft's performance, especially in larger propeller aircraft that tend to travel faster. A constant speed propeller automatically adjusts the blade angle. At low speed, the blade angle is kept small (also called fine pitch), therefore ensuring the angle of attack remains efficient for the RAF, as shown in Figure 9.8. At higher speeds, the blade angle is increased (also called coarse pitch), again to ensure the angle of attack remains efficient.

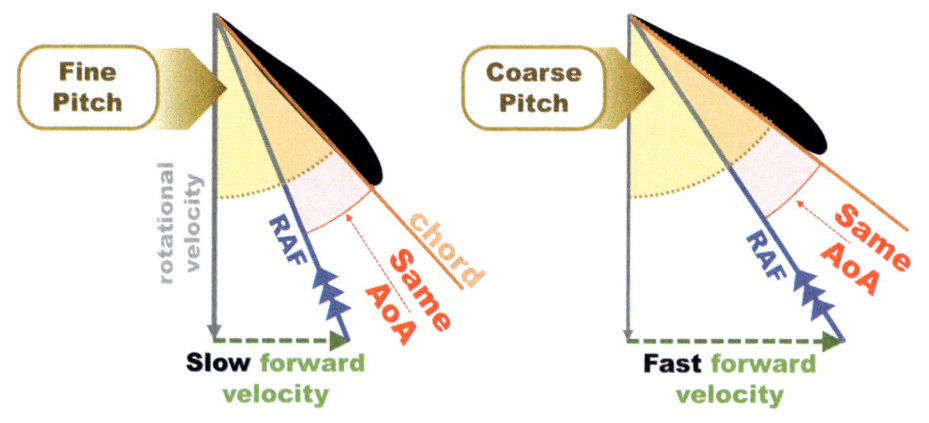

Figure 9.8: *A constant speed propeller automatically adjusts the blade angle.*

The inner workings of a constant speed propeller are quite amazing, with several different mechanisms used in aircraft to adjust the blade angle. One common type utilizes oil to adjust the angle hydraulically, as shown in Figure 9.9. Key components include a governor that moves oil back and forth through the propeller spinner, this is the same oil that was explored in an earlier chapter that is used to lubricate and cool the cylinders. The oil moves into an area with a piston that is connected to the propeller blade. Changes in oil levels result in the blade angle increasing or decreasing. There are stops that prevent the blade angle from changing too much in either direction and a spring that helps the propeller return to a low pitch setting.

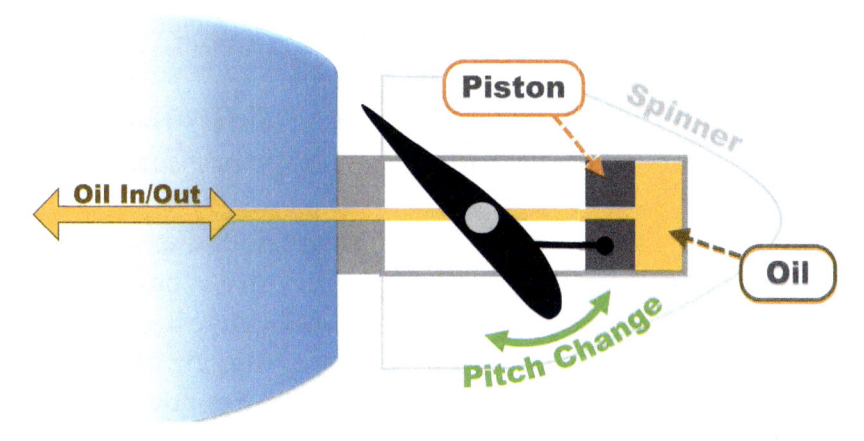

Figure 9.9: *Simplified components of a constant speed propeller.*

The blade angle is adjusted automatically when the pilot selects the engine settings in the cockpit. In an aircraft with a fixed pitch propeller, the pilot usually has two engine controls available: mixture and throttle, as shown in Figure 9.10. The engine and rpm are directly tied to the throttle setting. That is, increasing the throttle (more power), rpm will increase.

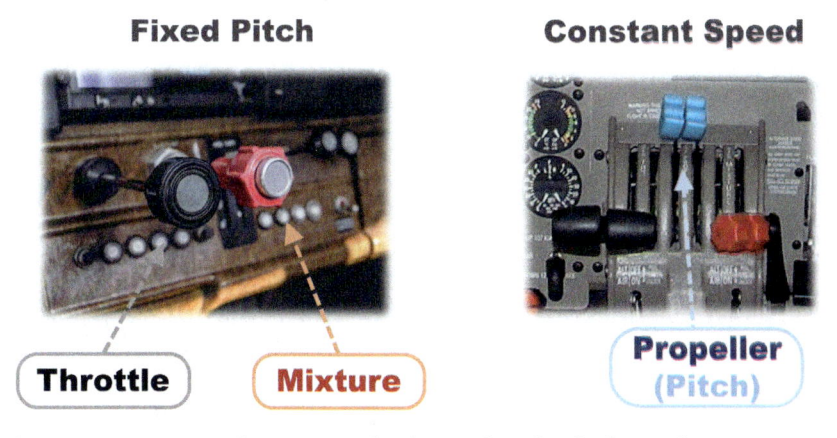

Figure 9.10: *Engine controls in a fixed pitch and constant speed propeller aircraft (note the constant speed controls are for a twin-engine aircraft, hence two levers for each engine control).*

Aircraft with a constant speed propeller will have an extra lever in the cockpit:

- The propeller lever (usually **blue**) controls the propeller rpm (also called pitch control).
- The throttle (usually **black**) controls the power of the engine by adjusting the manifold pressure (MAP).

Manifold pressure (MAP) is the pressure in the inlet manifold. MAP is basically how much fuel-air mixture is being delivered to the cylinders and is indicated on the MAP gauge in the cockpit. The higher the manifold pressure, the more fuel-air mixture is being delivered to the engine, and therefore higher power is produced. Pilots will usually select a specific rpm (e.g. 2,000 rpm) and adjust the MAP (with the throttle), for the desired power output. Any change in speed (e.g. aircraft descends) or MAP will result in a change in blade angle to maintain the selected rpm. The MAP gauge will have green and red markings, and for any given rpm, it is important pilots do not exceed manifold pressure limits. Exceeding these limits can place a huge strain on the cylinders, potentially leading to an engine failure.

Pilots need to be careful when changes are required to both the throttle (MAP) and propeller (rpm) settings, such as entering a descent or executing a go-around. The order that each control is moved is important to avoid overstressing the engine:

- When the power setting is *reduced* (e.g. during a descent), reduce the manifold pressure with the throttle first, and then reduce rpm with the propeller control. If rpm is reduced first, manifold pressure will automatically increase, possibly resulting in the manifold pressure exceeding limits.
- When the power setting is *increased* (e.g. during a go-around), reverse the power setting sequence. Increase rpm with the propeller control first, and then manifold pressure with the throttle.

Reduction Gearbox

In most small aircraft the propeller is directly connected to the engine's crankshaft (called a direct drive engine), meaning the propeller spins at the same speed as the engine. Most engines are designed to keep the propeller speed below 2,700 rpm. However, some engines have been designed to turn at very high speeds (e.g. 5,000 rpm). If the propeller was to spin at these fast speeds, the tips of the propeller could exceed the speed of sound, not only resulting in excessive noise but also a significant loss of thrust. To overcome this issue, an aircraft with an engine that operates at high speed will be fitted with a reduction gearbox. This clever device sits between the engine and the propeller, as shown in Figure 9.11, and can take the fast-moving engine and slow it down for the propeller – so effectively the engine spins faster than the propeller. Why not just have engines that can run at the same slow speed the propeller needs? Although this is the case for most small engines, the main advantage is some engines are more efficient when operating at high speed, therefore a reduction gearbox means both the engine and propeller can remain efficient.

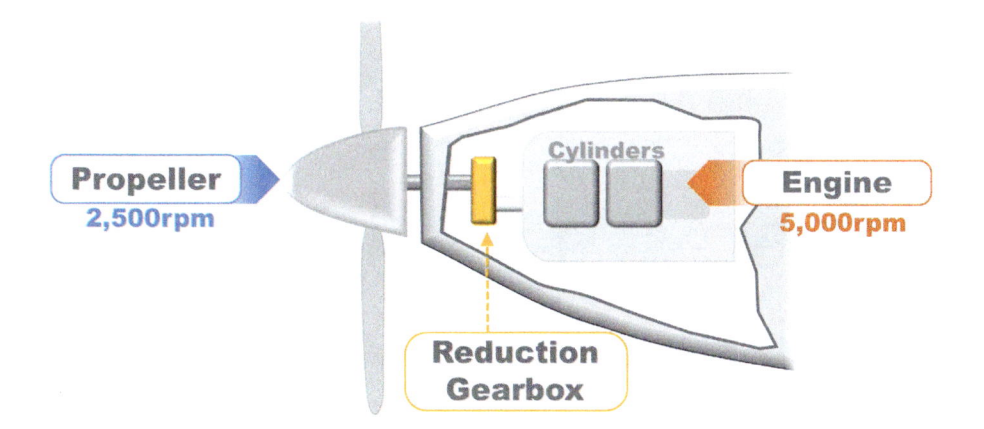

Figure 9.11: *A reduction gear box allows the propeller and engine to turn at different speeds, allowing both to operate efficiently.*

The propeller can seem daunting, but it is an essential component for flight. Just like the wing is required to generate lift to overcome the weight of the aircraft, the propeller is required to generate thrust to overcome drag. Before a flight, pilots should inspect the propeller to make sure it is fit for the flight ahead, such as looking out for any nicks or cracks. Pilots also need to treat the propeller and engine properly. Proper engine management techniques are an essential skill for all pilots, and will be explored in more detail in the next chapter.

CHAPTER 10: **ENGINE MANAGEMENT**

The last few chapters have explored the amazing parts of the aircraft engine, from the combustion of the fuel-air mixture inside the cylinders to the range of engine cooling systems. These different systems work effortlessly together, ensuring smooth engine operations, but they are also reliant on each other. For example, a problem in the cooling system can quickly result in the engine overheating, leading to an engine failure. In this chapter you will explore some general engine management techniques to help you look after the aircraft engine. Each engine and aircraft combination are unique; therefore, you must consult your Pilot's Operating Handbook for the specifics of the aircraft you operate.

Preflight Inspection

The first engine management technique occurs before the engine even starts, with the preflight inspection. A preflight inspection should be performed before each flight, inspecting various parts of the aircraft, including the engine. The importance of conducting a thorough preflight inspection is highlighted in the following case study:

> On the 13th of May 2016, a pilot was preparing for a flight from an aerodrome in Tennessee, in a single-engine Piper PA22. During the preflight inspection, the pilot found three baby birds in the cockpit. After removing the birds, the pilot continued the inspection. Shortly after take-off, a fire started in the engine compartment and smoke began pouring out of the engine cowling. The pilot returned to the aerodrome,

but due to the smoke limiting the forward visibility, the pilot lost control of the aircraft on landing. Everyone managed to escape the aircraft safely before it was engulfed in flames. The post-accident investigation found the remains of a bird's nest inside the engine cowling by the exhaust manifold, which was the likely origin of the fire.

What you inspect in the engine and how much access you have to it will vary between aircraft. For example, in some aircraft you can easily open the top part of the cowling, allowing you to have a really good look around the engine, whereas others have very limited access. In any case, key areas to inspect include checking the air inlet behind the propeller is clear (for such things as a bird's nest). If you can inspect inside the engine cowling, look for any loose connections (e.g. oil lines, fuel lines, high tension leads) and any signs of leaked oil. Oil on the bottom of the cowling may be an indication of a broken seal. Always check the oil levels are correct and top up if required (with the correct type of oil).

Starting the Engine

With the preflight inspection complete, it's time to start the engine. However, before starting it is essential you have a good look around to ensure no one is near the aircraft. It is the pilot's responsibility to ensure the area is clear before starting. Also, make sure the parking brake is set. Several accidents have occurred when an aircraft has unintentionally taxied straight into a nearby building or aircraft.

With the area all clear, it's time to start the engine. Ideally, the outside air temperature is mild, and the engine has not been run for the last couple of hours. You first set the power controls, engage the starter and the engine bursts into life. Unfortunately, engine starts do not always go this smoothly. This may sound a little odd, as when you jump into a car it is rare to worry about starting the engine, even if it is freezing outside. But this is largely due to cars having clever electrical starting systems –

something that's rare in small aircraft. Aircraft can sometimes be very temperamental, especially when it's cold or hot.

A **cold start** means the engine is cold, also known as cold soaked. This can be expected on the first engine start of the day, especially if it is cold outside. The engine often needs a little helping hand to start. The cold start procedure will vary between aircraft but usually includes priming. Priming means squirting a small amount of fuel directly into or just before one or more cylinders, providing an initial charge of fuel. A small amount of priming is used during most starts, but when it is cold, the maximum amount of priming should be applied (the priming system will be covered in more detail in the next chapter). In very cold temperatures extra measures may be required. This may include prewarming the engine or rotating the propeller (and therefore the engine) a few times to free up sticking oil seals. Pilots should always be very careful handling the propeller and should only do so after receiving proper training. The ignition switch must be in the OFF position, to ensure the magnetos are off when handling the propeller.

A **hot start** means the engine has run very recently (generally in the last two hours) and is less concerned with the outside air temperature, but the temperature of the engine. Hot starts are usually more challenging with an engine with a fuel injection system. This is because after the engine is shut down, some fuel remains in the fuel lines inside the engine cowling. The area inside the engine cowling remains very warm, resulting in the fuel vaporizing inside the fuel lines – known as vapor locking. This can cause a blockage that can persist until the fuel lines cool down. Hot start procedures will often involve operating the fuel boost pump (electric fuel pump) which pressurizes the fuel lines, helping to remove the vapor in the system.

A **flooded** start may occur when there is too much fuel in the cylinders when starting the engine, such as priming for too long (over-primed engine). The fuel in the cylinders will struggle to ignite, as it is too rich. The general procedure is to leave the mixture control in the idle cut-off position (so no more fuel enters the cylinders), engage the starter, and once the engine starts, advance the mixture control to full rich.

Start procedures vary considerably, therefore you must be familiar with the procedures for your aircraft. Hand-swinging a propeller is sometimes used as a means of starting an engine, but this must only be conducted with extreme care, and with proper training. Always treat the propeller as live, meaning any time you are walking near the propeller (e.g. during a preflight inspection), position yourself away from the area the propeller would rotate if it suddenly came to life

After Start

Once the engine has started, you should begin your regular checks of the engine instruments. As discussed in various chapters, these range from oil temperature and pressure, to cylinder head temperature and manifold pressure. After the engine has started, oil pressure should sit within the normal range, but oil temperature may take a little longer to respond, especially after a cold start. Engine instruments should be monitored and cross-checked together, as an abnormal indication in one will often be associated with an abnormal indication in another. For example, low oil pressure is often associated with high oil temperature. When an abnormal indication occurs, one of the easiest responses is to assume it's a gauge or sensor error. Although this will occur occasionally, it's always best to play it safe and treat it as if the gauge or sensor is working correctly.

Prior to take-off, you should have conducted the normal engine checks, which will usually include checking the magneto system. On the ground, you should also be careful of the surface you are taxiing over. For example, taxiing over a stony surface can result in stones being picked up and thrown into the propeller. A nick on the propeller can lead to a crack and eventual blade failure if it goes unnoticed. When you are ready to take-off, this is your first opportunity to use the throttle through its full range. *Advance the throttle smoothly.* As a general rule, it should take about three seconds from idle to full throttle position. This same principle applies when reducing power (e.g. to begin a descent). Misusing the engine controls can lead to engine damage.

Rough Running

Imagine flying along on a long cross-country flight when the engine begins to run rough. You will likely notice the rough running immediately and it may send you into a mini panic. But it is important to stay calm and run through the engine rough running procedures for your aircraft, which may include:

Check the power settings, especially the mixture control. If the fuel-air mixture is too lean, the engine will run rough. Place the mixture into the full rich position.

Check the fuel selector. You will cover fuel in more detail in the next chapter, where you will see small aircraft often have multiple fuel tanks (e.g. left tank, right tank). You may be running low on fuel on one tank and can select another.

Turn on the electric fuel pump. Some aircraft (e.g. low wing or aircraft with a fuel injection system), will have an engine-driven fuel pump to deliver fuel to the engine. If the engine-driven fuel pump fails, you can turn on the backup electric fuel pump (also called a boost or auxiliary pump).

Is the primer on? As touched on earlier, the primer squirts raw fuel into or near the cylinder(s). The primer should not be used during normal operations, as this will result in the wrong fuel-air ratio entering the cylinders (too rich). Some aircraft have a manual primer that should be locked.

If you are flying an aircraft with a carburetor, *carburetor ice* may be the cause of the rough running engine. You may recall from chapter six, carburetor ice can occur even on a warm day, provided there is sufficient moisture in the air. Select carburetor heat, noting that an initial reduction in power is normal due to the warm air and possible melting ice entering the cylinders. As the ice melts, the power should increase.

Faulty ignition system? Using the ignition switch, select the left and right magnetos. If the engine runs smoother on one, and not BOTH, then select the single magneto that gives the smoother running. Be aware, you are now only operating on a single magneto system, and if a fault occurs with this one system, a total engine failure is possible.

If the engine continues to run rough, divert to the nearest aerodrome. If you have insufficient power to remain in level flight, you may need to consider conducting an off-field landing.

Full Authority Digital Engine Control (FADEC)

If you are familiar with the engine technology in cars, you might be forgiven for thinking aviation is considerably behind in terms of engine technology. But there is a good reason for this, as the engine systems found in aircraft are reliable, with various backup systems available in case something fails. However, small aircraft engines are beginning to slowly embrace advances in technology in the form of **full authority digital engine control (FADEC)**. FADEC has been used in large aircraft for a considerable time but is starting to appear in small aircraft too.

FADEC contains a digital computer that controls the engine and the propeller, largely simplifying engine management. If you jumped into an aircraft with FADEC, the first thing you will notice is there is just one engine control (replacing the throttle, mixture and propeller controls). The next thing you will notice is how the system significantly reduces your engine management workload. The system has multiple sensors around the engine so it can digitally control all parts of engine operations; from spark timing, fuel-air mixture, priming and injector timing. The pilot simply positions the single lever to the desired position, such as start or cruise power, and the FADEC system automatically adjusts the engine and propeller. Such control means reduced fuel burn and more power. What is the downside? The main problem is if the FADEC fails, so does the engine. This would include loss of electrical supply to the FADEC system. To protect from such a problem, a FADEC will usually have a redundant FADEC (i.e. a duplicate system that can kick in if one fails), and also a backup electrical power supply.

The need for a backup electrical power supply was highlighted in an unusual accident in Germany in 2007 of a small, twin-engine Dimond DA42. At the time the DA42 was a relatively new aircraft, which had digital engine controls. During preparation for a flight, the pilot found the aircraft's battery was flat, therefore started both engines with a ground power unit. This was against the procedure stated in the Pilot's Operating Handbook, which stated only one engine should be started with the ground power unit, with the second engine started using aircraft-generated power (e.g. electricity stored in the aircraft battery). This procedure was designed to confirm the aircraft could handle the electrical load placed on it during a flight. Shortly after take-off the pilot raised the undercarriage, causing a very short-term drop in the electrical voltage to the two engine control units (ECU). This extra load would normally be covered by the battery, however, in this case, the battery was still flat. This resulted in both ECUs going offline, stopping both engines. With very little altitude, the pilot made a successful belly landing into a field next to the runway. The accident highlighted the importance of backup systems, to cover a range of potential issues that can affect the FADEC system.

Pilots need to treat their engine with respect. If the engine is treated poorly, there is a greater risk of it failing, which can have grave consequences. Understanding the theory side of the engine is essential for all pilots, giving them the confidence to make the right decision when things are not running smoothly. You have spent the last couple of chapters exploring the various parts of the engine, next you will look at one of the components required to keep the engine running – fuel.

CHAPTER 11: **FUEL**

Even small aircraft will carry a considerable amount of fuel, allowing them to travel long distances. This fuel needs to be stored safely in the aircraft and delivered in a continuous flow to the engine. However, ensuring the aircraft engine has a steady supply of fuel is not an easy task. Aircraft operate over a wide range of flight conditions; from extreme flight attitudes (e.g. pitching up sharply) to high altitudes where the temperature and pressure are much lower than sea level. Pilots also need to carefully manage the fuel in the aircraft, with poor fuel management one of the leading causes of aircraft accidents. Fuel management starts well before an aircraft departs, as pilots must ensure they have the correct fuel, enough of it, and the fuel is free of contaminants (e.g. dirt, water). This chapter begins by exploring the different fuel types pilots may encounter, followed by a look at the typical fuel systems found in small aircraft.

Fuel Types

When an aircraft is refueled, there may be several fuel types available. Selecting the wrong type of fuel could have dire consequences. **Aviation gasoline (AVGAS)** is the most common type of fuel used in small aircraft and is identified by an octane or performance number. The number relates to the **anti-detonation** qualities of the fuel. The higher the number, the higher the pressure the fuel can withstand before detonation. The most common type of AVGAS used in small aircraft is called

100 or 100 LL (rating 100, low lead). Lead is added to AVGAS to improve its anti-detonation qualities. AVGAS with low lead (LL) has a reduced amount of lead, with other chemicals added to maintain the anti-detonation qualities. Lower octane fuels may also be available, such as AVGAS 80 (80 octane rating). These may be used in engines with lower compression, as the fuel ignites at a lower pressure. The Pilot's Operating Handbook will state the fuel required for the engine. Never use a grade lower than recommended as this can lead to detonation and engine failure. Automobile gas, also called **MOGAS,** should not be used in the aircraft unless the aircraft and engine are certified to use it. MOGAS may not be at the high-quality standard required in aviation and will usually have different burning characteristics to AVGAS.

The other type of fuel used in aviation is called **jet fuel** or **AVTUR.** This is primarily used in turbine-powered aircraft (e.g. turboprops), but is also used in some small aircraft (e.g. with a diesel engine). The correct type of fuel must be used in the aircraft, as highlighted in the following case study:

> On the 5th of October 2019, a twin-engine Piper Aerostar 602 was preparing to take-off from an aerodrome in Indiana. The pilot had requested fuel from a fuel truck, and when it arrived the refueler checked with the pilot that he wanted *'jet fuel',* to which the pilot replied *'yes'*. Initially, the refueler had trouble fitting the fuel nozzle into the aircraft's fuel tank, but by angling the fuel nozzle by about 45 degrees, the aircraft was refueled with about 163 gallons of Jet A fuel (Jet A is a common type of jet fuel). The aircraft started up fine and took off, however, shortly afterward both engines failed, and the aircraft crashed into the ground. The cause of the engine failure was the aircraft was refueled with the wrong fuel. The aircraft's engines required AVGAS (100LL), rather than the jet fuel that was used to fill up the aircraft. The wrong fuel in the engine will lead to detonation in the cylinders

which can lead to catastrophic engine failure. To reduce the risk of the wrong fuel being placed into an aircraft, the fuel nozzles for specific fuel types are different. This is why the refueler in this case was having trouble fitting the nozzle into the aircraft's fuel tank.

Due to the importance of using the correct fuel, each fuel is designated a unique color, which is used both in the fuel itself and also on part of the fuel labeling:

- **AVGAS 80 = Red**
- **AVGAS 100 = Green**
- **AVGAS 100LL = Blue**
- **JET A = Black** (colorless or straw fuel color)

Figure 11.1: *Examples of fuel color labeling (the left side of the label indicates the fuel color, e.g. blue for AVGAS 100LL).*

Fuel Contamination

One of the important tasks pilots complete during a preflight inspection is to take a sample of fuel. This simple task is to look for anything that is lurking in the fuel that shouldn't be there, such as dirt or water. If these contaminants make their way to the engine, at best they will cause a short patch of engine rough running, at worse, an engine failure. Each aircraft will have easily accessible fuel drains, often several of them. These are located at low points in the fuel system, such as at the bottom of each fuel tank. Water is the most common type of contaminant, and as it is heavier than fuel, will naturally collect at the low points where the samples are taken. Water in a fuel sample can be identified as either a cloudy appearance or a clear separation between the water and the fuel, as shown in Figure 11.2. The sample can also be checked to make sure the correct fuel type has been placed in the aircraft (e.g. checking the fuel color).

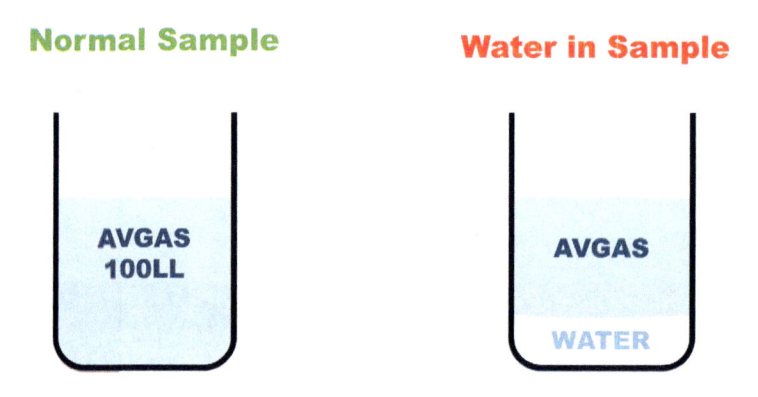

Figure 11.2: *Water is the most common type of contaminant in fuel and will sink to the bottom of a fuel sample.*

If contaminants are found in a sample, further samples should be taken until they no longer appear. If after repeated samples the contaminants remain, the aircraft should not be flown. To reduce the risk of water entering the fuel tanks at night, pilots can consider filling up the fuel tanks after the last flight of the day. This reduces the risk of condensation forming in the empty parts of a fuel tank when the temperature is cooler. The downside of filling up the fuel tanks is this may cause weight issues for the next flight (e.g. aircraft is above maximum take-off weight with full fuel tanks) and if the temperature increases, the fuel may actually expand and leak out of the fuel tanks.

Extra care should be taken if refueling from fuel cans or drums. The quality of the fuel is harder to determine, as shown in the following case study:

> On the 18th of July 2010, a twin-engine Aero Commander was preparing to take-off from a remote aerodrome in Northern Canada. No fuel trucks were available at the remote aerodrome, therefore two 45-gallon drums of AVGAS were delivered to the aircraft from a local fuel supplier. The pilot used all of one drum, and most of the second to fuel the aircraft. Prior to take-off all engine indications were normal, however shortly after becoming airborne, the pilot observed a rapid increase in cylinder head temperature in both engines before losing power. The aircraft was unable to maintain altitude, therefore the pilot made a successful forced landing nearby. The engine failure was traced to the 'fuel' found in the second 45-gallon drum. Both drums looked very similar and were marked AVGAS 100LL. However, the second drum contained a mixture of left-over fuels (diesel, jet fuel) as well as various contaminants like dirt. This drum should not have been delivered to the aircraft but had been placed with other AVGAS drums at the fuel supplier. The pilot checked the first drum and found it contained the correct type of fuel and assumed the second drum was the same. The incorrect fuel mixture caused the cylinders to overheat, leading to detonation and engine failure.

Fuel System

Fuel is stored in several tanks in the aircraft and needs to be delivered to the engine under all normal flight conditions. There are two main fuel systems found in small aircraft: **gravity-fed** and **fuel pump** systems. A gravity-fed system uses gravity to deliver the fuel to the engine, such as from a fuel tank in the wing of a high wing aircraft. This is a fairly simple system, suitable to deliver fuel via a carburetor, as shown in Figure 11.3.

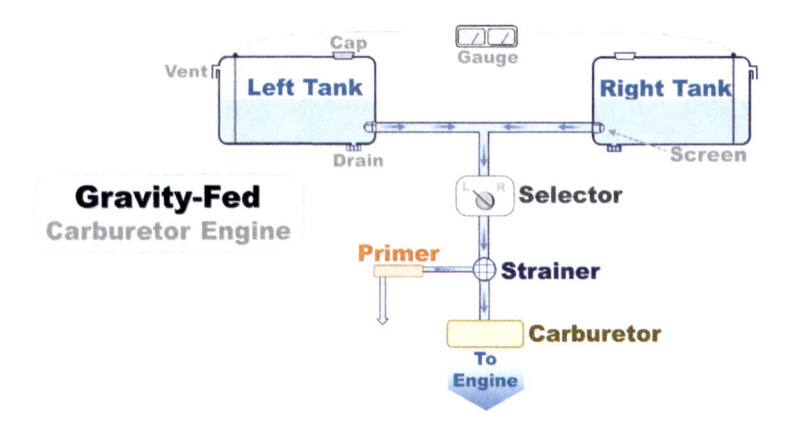

Figure 11.3: *Typical gravity-fed fuel system (with a carburetor engine).*

A gravity-fed system is less suitable in a low wing aircraft or an aircraft with a fuel injection system. In these aircraft, fuel pumps are used to deliver fuel under pressure to the engine, as shown in Figure 11.4.

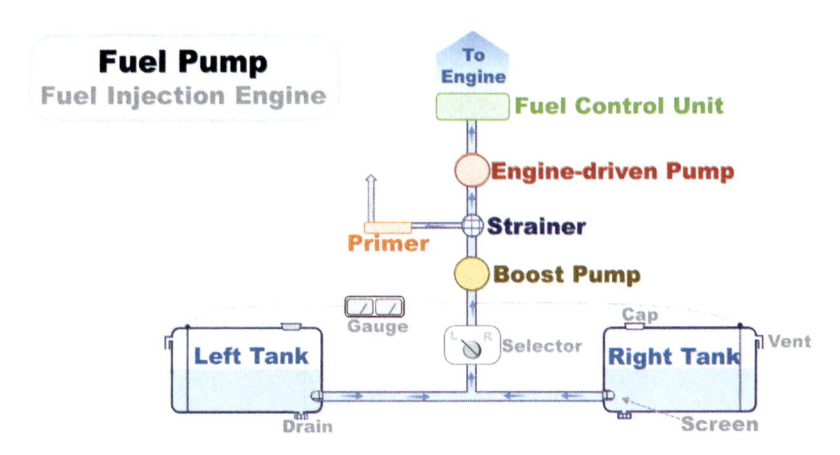

Figure 11.4: *Typical fuel pump system (fuel injection engine).*

Let's go through the journey of the fuel, starting at the fuel tank. **Fuel tanks** are normally located inside each wing, with a filler cap on top for refueling. Each fuel tank has a **fuel vent** to ensure the air pressure above the fuel is the same as outside (if the vent is blocked, fuel may struggle to flow out of the tank). A tank will also have an **overflow vent**. Pilots may see fuel dripping out of the overflow vent after filling up on a warm day, due to the fuel expanding in warmer temperatures. A **sample drain** will be located at the low point in the fuel tank, the same drain discussed earlier to collect fuel samples. Some fuel tanks may have multiple fuel drains. A **screen** or filter is fitted to the fuel outlet to help prevent contaminants from leaving the fuel tank. To further protect from contaminants entering the fuel system, the fuel line leaving the tank does not sit right at the bottom (where contaminants are more likely to collect). However, the raised fuel outlet does mean there is usually a small amount of unusable fuel at the bottom of each tank.

The **fuel gauge** in the cockpit will indicate how much fuel is currently in each tank. In most small aircraft these are electrically operated, therefore the aircraft's electrical master switch must be turned on for the gauge to display a fuel reading. Fuel gauges can be unreliable; therefore it is essential pilots visually check the fuel levels during the preflight inspection, often utilizing a fuel dipstick.

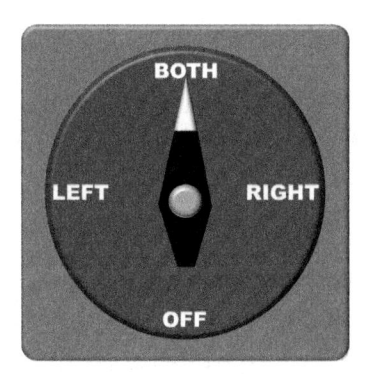

Figure 11.5: *Example of a typical fuel selector.*

After leaving the fuel tank, fuel will then usually pass through the

fuel selector valve, which is controlled in the cockpit. The fuel selector allows the pilot to select fuel from a particular tank, as shown in Figure 11.5. Common selector positions include: LEFT, RIGHT, BOTH and OFF. The LEFT and RIGHT positions will take fuel from only that particular tank (e.g. when pilots select LEFT, only fuel from the left tank will be used). When using fuel from one tank, pilots should avoid running the tank dry. Running a tank dry could result in air entering the fuel system, leading to engine problems, as shown in the following case study:

> On the 18th of July 2011, a pilot was conducting a sightseeing flight over Quebec, Canada, in a float-equipped single-engine Cessna 185. The aircraft had two fuel tanks (left and right), with sufficient fuel on board to complete the planned flight. It was normal procedure to have the fuel selector on BOTH during critical phases of a flight (e.g. take-off), but in the cruise, tank selection was at the pilot's discretion. It was common practice for pilots to select the left fuel tank during the cruise, as this made it easier to refuel (due to difficulties accessing the right tank when the aircraft was at the dock). Unfortunately, the pilot misjudged the amount of fuel in the left tank, and the engine lost power due to fuel starvation. The pilot was unable to restart the engine despite plenty of fuel in the right tank and the aircraft crashed into a river below.

An aircraft with a **fuel pump** system will need a pump to deliver fuel to the engine (a gravity-fed system does not need a fuel pump). The system will have two pumps: an engine-driven and an auxiliary fuel pump:

- The **engine-driven** fuel pump is the main pump, driven by the engine and usually operates automatically (e.g. no on/off switch in the cockpit).
- The **auxiliary pump** is electrically driven and used when starting the engine. It is also used as a backup in case the engine-driven fuel pump fails.

In some aircraft the auxiliary fuel pump may be left on during critical phases of flight, such as take-off. This pump may also be called a boost pump. A fuel pump system will also have a fuel pressure indicator in the cockpit.

The fuel will pass through a **strainer** at some point in the fuel system. The strainer is designed to collect any moisture and other contaminants in the fuel. Aircraft may also have a **sump,** which is a low point in the fuel system, which may be located with the fuel strainer. This point in the system will also contain a drain that a fuel sample should be collected from (just like the wing drains).

The **primer** is usually used during engine starts, especially cold starts, as discussed in the previous chapter. The primer may be a simple hand-operated pump or an electrically driven pump. The primer pulls fuel from the main fuel line (usually at the fuel strainer) and injects a fine mist of fuel directly into the induction manifold near one or more cylinders (some may spray directly into the cylinder itself). It is important the primer is locked (or turned off) when not being used, otherwise the fuel-air mixture will be too rich.

Refueling

When refueling an aircraft, it is important to take the necessary steps to reduce the risk of static electricity igniting the fuel. It is common for different objects to generate different levels of static electricity. The problem occurs when two objects with different charges meet, potentially causing a spark. Not surprisingly a spark during refueling is not ideal, potentially igniting the fuel vapor in the fuel tank. To protect against static electricity igniting the fuel, the aircraft should be grounded or bonded to the refueling equipment (also known as earthing). Bonding wire is used to connect the refueling equipment to the aircraft, which equalizes the charge. The bonding wire must be attached to bare metal on the aircraft, such as the exhaust pipe (careful, this may be very hot if the aircraft has just been flown). When refueling is complete, the reverse process should be followed. That is, place the fuel nozzle back into its correct position before removing the bonding wire. You should always take a fuel sample after refueling, however, it is important to wait for the new fuel to settle. This will allow any contaminants to sink to the bottom of the tank before the fuel sample is collected.

Even small aircraft carry a considerable amount of fuel. The fuel needs to be delivered in a continuous flow to allow the engine to run smoothly over a range of flight conditions. But like many of the other systems explored in this book, the fuel system is only as good as the pilot operating it. Poor fuel management can have disastrous consequences, with numerous accidents occurring as a result.

CHAPTER 12: **ELECTRICAL SYSTEM**

Imagine flying along when the aircraft's cabin lights begin to dim, followed a short time later by a full electrical systems failure. Is the aircraft about to plummet to the ground? Unlikely, most aircraft are designed to ensure the engine runs perfectly fine after a total electrical failure, however, many systems will be affected, ranging from the radio to the landing light. The electrical problem may be fixed easily. But to troubleshoot an electrical problem, it is important that pilots have a general understanding of the electrical system in their aircraft.

When the aircraft's electrical system is switched on, it supplies electricity to a range of equipment, which may include:

- Radio equipment
- External lights (e.g. landing, position and anticollision lights)
- Internal cabin lights
- Some instruments and gauges (e.g. fuel gauge)
- Electric fuel pump
- Pitot heat
- Starter motor
- Flaps (if operated electrically)

A typical small aircraft will have a relatively simple, **direct current (DC)** electrical system of 14 or 28 volts. Direct current means the electrical current (amperes or amps) flows only in one direction. You may also hear some aircraft have an alternating current

(AC) electrical system, whereby the electrical current reverses direction. **Volts** are basically the size of the force that sends the electrical current.

Let's start by exploring the key components of a basic electrical system, as shown in Figure 12.1.

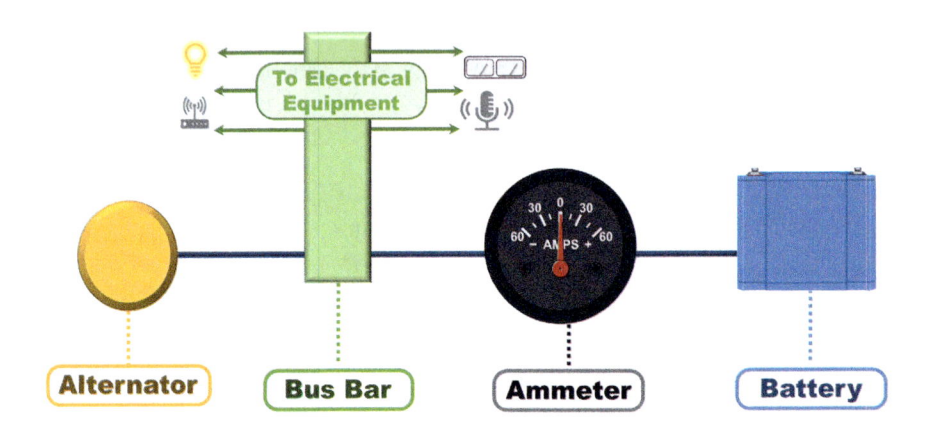

Figure 12.1: *Basic components of a small aircraft electrical system.*

Battery: A battery is used to store electrical power. The battery will supply electrical power to start the engine and can provide a short-term backup in flight. However, the battery will quickly run out of electrical power if it is not recharged.

Alternator/Generator: An alternator or generator can create electrical power when the engine is running. They can recharge the battery and provide a continuous supply of electrical power throughout the flight. An engine-driven alternator is more likely to be found in modern small aircraft as they are reliable, especially at low engine power settings (e.g. when the aircraft is on the ground) and are relatively small. The main downside of an alternator is it requires an initial current from the battery before it can produce electrical power. This means if the aircraft is started without the battery, the alternator will not come online. A generator may be found on older aircraft, but they tend to be larger and heavier, and sometimes have trouble generating sufficient electrical power at low engine power settings. The main advantage of a generator is it does not need a battery to start producing electrical power.

Master Switch: Many small aircraft will have a split-type master switch (especially one with an alternator) that controls all the electrical power to the aircraft. One side of the switch controls the battery, whereas the other half controls the alternator, as shown in Figure 12.2. This means the pilot can turn off the alternator (e.g. if it is overcharging) but the electrical system will continue to operate through the battery (which will quickly deplete). Both switches are normally in the ON position during a flight.

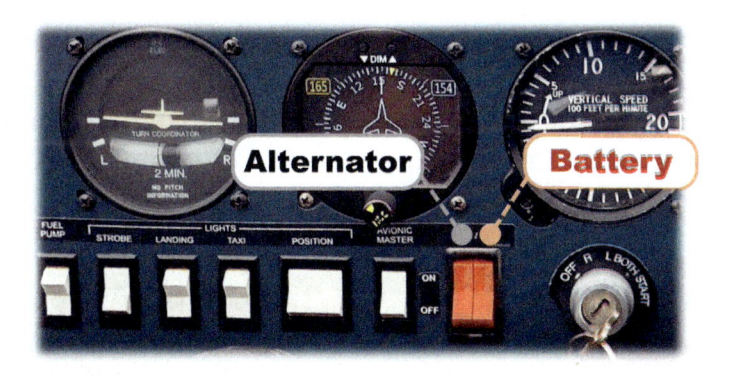

Figure 12.2: *The master switch.*

Bus Bar: A bus bar is used to greatly simplify the electrical wiring of an electrical system by providing a common point that the electrical power is sent to, and then distributed to a range of electrical equipment in the aircraft.

Circuit breakers/fuses: Circuit breakers or fuses are designed to protect the aircraft wiring and electrical equipment. They work by opening the circuit (cutting off the flow of electricity) in the event of an overload. They are usually designed to trip (pop) at a slightly lower rating than the associated wiring, meaning they will trip before the rating of the wire is exceeded and the wire overheats. A circuit breaker can be manually reset, whereas a fuse needs to be completely replaced (as an overload normally causes a fuse wire to melt). Extreme caution is required when resetting circuit breakers or fuses, which will be discussed in more detail shortly.

Ammeter: An ammeter is used to show if the alternate/generator is supplying sufficient electrical current to the electrical system. There are two main types of ammeters: a left-zero ammeter or a center-zero ammeter, as shown in Figure 12.3. A left-zero ammeter only measures the output of the alternator or generator, with zero on the left, and may also be called a loadmeter. During normal operations, it should be reading slightly above zero. A center-zero ammeter measures the electrical current in or out of the battery. A positive indication means the battery is being charged, whereas a center indication means the generator/alternator is matching the electrical discharge from the battery.

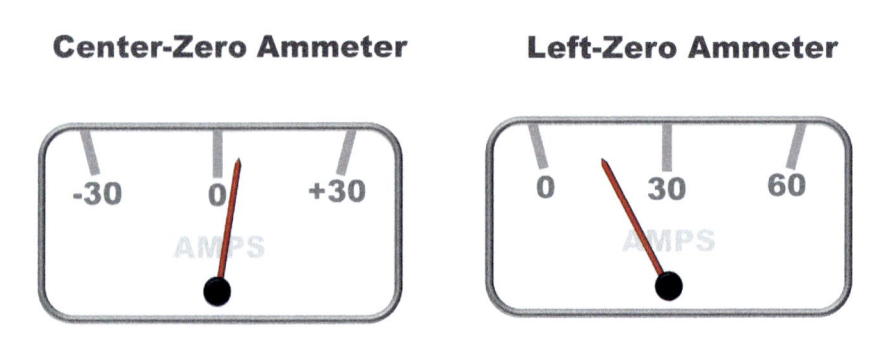

Figure 12.3: *Examples of the two main types of ammeters.*

Monitoring the Electrical System

It is important pilots look after the health of their aircraft's electrical system, beginning before the engine even starts. Pilots should limit the use of electrical equipment before the engine is running. During the preflight inspection of the aircraft, pilots may need to check some electrical equipment, such as external lights, but this should be kept to a minimum. The next major drain on the battery is starting the engine. When the electrical starter is engaged, it draws a large electrical current from the battery. If you are having trouble starting the engine (e.g. cold start), you may drain the battery. If the battery is depleted before the engine starts, a ground power unit may be able to provide electrical power to get the engine started.

Once the engine has started, the alternator/generator will begin to supply electrical power to the aircraft, and the ammeter is one of the key tools for monitoring the health of the electrical system:

- A left-zero ammeter should display a positive indication at all times. If it displays zero, then the alternator/generator is not supplying any electrical power – and the battery is the only source.
- If a center-zero ammeter displays a negative reading, then more current is leaving the battery than is being replaced, again, an indication of a malfunction of the alternator/generator.

In the event you are solely relying on the battery, it is important to be aware it will only be able to supply sufficient electrical power for a short period of time. As a result, it is important to turn off non-essential electrical equipment and be prepared for a full electrical failure. For example, you may be able to warn air traffic control that you may be losing your radios shortly. As noted earlier, pilots operating an aircraft with a generator should keep an eye out for insufficient charge during low engine power settings. You should also be on alert for overcharging. It is common to observe a large charge indication on the ammeter shortly after the engine starts (as the battery needs recharging), but this should settle down after a short period (normally within 30 minutes). If the excess charge continues, it can cause the battery to overheat. Overcharging would normally result in the need to turn off the alternator/generator and follow the above procedures of relying on the battery alone.

A damaged or overheated electrical system can lead to an electrical fire. Although rare, an electrical fire in flight is an extremely dangerous situation. The first signs of an electrical fire are normally very subtle, such as a slight burning odor or tripped circuit breaker. Once the fire starts, pilots will have very little time to respond, as shown in the following case study:

> On the 10th of July 2007, two pilots took off in a twin-engine Cessna 310 from an aerodrome in Florida. Shortly after leveling off in the cruise, about 10 minutes after departure, the pilots declared an emergency and reported smoke in the cockpit. Air traffic control cleared the aircraft direct to a nearby aerodrome. A few seconds later, the controller heard a partial radio call *'shutoff all radios, electrical….'*, which is consistent with the pilots following the inflight fire or smoke procedure. A short time later the aircraft was observed flying very low and fast, before crashing into a residential area. The day before the accident, another pilot flying the same aircraft experienced a problem with the weather radar, including noticing a burning smell. The pilot pulled the circuit breaker for the weather radar, which resulted in the burning smell going away and continued to fly the aircraft without any other problems. The pilot reported the issue on return, but it was not investigated further, nor was a collar placed around the circuit breaker to stop it from being reset. The accident pilot was briefed about the weather radar issue; however, it is likely the circuit breaker was reset, reenergizing the faulty wiring, leading to the inflight fire.

Pilots should be very cautious about resetting a tripped circuit breaker (either before a flight or in the air). Small aircraft will usually fly perfectly fine without the electrical system, therefore very few circuit breakers need to be reset to ensure a flight can be completed safely. Historically pilots have been taught it is fine to reset a tripped circuit breaker once as the problem may be intermittent or a one-off, and resetting the circuit breaker will restore the electrical component. This practice is based on the assumption that if the problem remains, the circuit breaker will simply trip again, and therefore the aircraft will be protected from any electrical faults. Unfortunately, this is not always the case, and several high-profile accidents have shown the dangers of resetting circuit breakers without fully understanding why they popped in the first place. If a circuit breaker does need to be reset (e.g. essential for the flight), first check there is no smell or other signs of burning or overheating. If the circuit breaker pops again, do not reset it a second time.

The aircraft's electrical system helps keep many components in the aircraft operating – from lights to some instruments. However, it is important that pilots remain alert for a faulty electrical system, which can be very dangerous. Aircraft are usually designed to operate fine without the electrical system; therefore pilots should not hesitate to turn off electrical equipment if there is a concern about the health of the system.

CONCLUSION

The design features that allow an aircraft to fly are amazing. The main airframe structures are strong yet remain as light as possible to allow the aircraft to leap into the air. The engine is very reliable, despite having many fast-moving parts and producing vast amounts of heat. The numerous aircraft supporting systems quietly operate; from helping to keep the lights on to providing a continuous supply of fuel. Despite aircraft being resilient, pilots need to treat them with respect. Poor engine handling, insufficient oil, or inadequate fuel management can all have dire consequences.

We started our aircraft technical knowledge journey by exploring the main airframe structures. These structures are designed to withstand the various flight loads an aircraft may be exposed to during a flight. These structures have evolved with advancements in technologies, including composite materials that are now common in small aircraft. Next, we explored some of the smaller aircraft structures; the flight controls and the undercarriage. Despite being small, they play an important role during various phases of a flight. We then took a look at the aircraft engine. The engines found in small aircraft come in many shapes and sizes. They can produce a considerable amount of thrust to help an aircraft overcome drag. Then we explored the range of supporting systems to help keep the engine running smoothly, from removing the toxic exhaust gases to cooling systems to avoid overheating. The final chapter explored the aircraft electrical system. We saw that this system supplies electrical power to many components in the aircraft, but pilots need to remain on alert, as a faulty electrical system can be very dangerous.

Aviation is fast evolving, with even small aircraft becoming more complex. Despite aircraft coming in a range of sizes, many have the same basic components. Whether you are a pilot who is just experiencing the joy of flying for the first time or flying has been your passion for countless years, it is essential you have a solid understanding of aircraft technical knowledge.

INDEX

Aileron, 15
Alternator, 102
Ammeter, 104
Carburetor, 47
Carburetor ice, 53
Circuit breakers, 103
Composite, 9
Compression ignition (Diesel) engine, 38
Constant speed propeller, 78
Cooling systems, 66
Connecting rod, 31
Cowl flaps, 69
Crankshaft, 31
Cylinder, 31
Detonation, 37
Electric starter system, 59
Electrical system, 101
Elevator, 16
Empennage (tail section), 13
Exhaust manifold, 42
Exhaust System, 41
Fixed pitch propeller, 76
Flaps, 21
Four-stroke cycle, 33
Fuel contamination, 93
Fuel injection, 54
Fuel system, 95
Fuel types, 91
Full Authority Digital Engine Control (FADEC), 88
Fuselage, 5
Impulse coupling, 59
Inlet and exhaust valves, 31, 35
Lubrication (oil) system, 62
Manifold pressure, 80

Magneto ignition, 56

Oil, 61

Oil pressure and temperature, 64

Piston, 31

Pre-ignition, 37

Propellers, 71

Reduction gearbox, 81

Refueling, 99

Rough engine running, 87

Rudder, 17

Spark ignition engine , 31

Spark plugs , 31

Starting the engine, 84

Trim tab, 23

Two-stroke engine, 39

Undercarriage (landing gear), 25

Viscosity, 61

Wet sump system, 62

Wing, 10

BOOKS IN THIS SERIES

Book 1: Human Factors for the Private Pilot

From the dangers a pilot faces when straying too high in an oxygen-deprived atmosphere to the way the brain attempts to process the enormous amount of information obtained during a flight. This book is for pilots and non-pilots to explore the vast number of factors that can influence a pilot's ability to fly an aircraft safely.

Book 2: Aviation Weather for the Private Pilot

Aviation weather is a wondrous and frightening subject. Pilots can encounter a range of weather conditions on just a single flight, from a towering thunderstorm that can toss an aircraft around like it is in a washing machine, to thick fog in which pilots will struggle to see just a few feet in front of the aircraft.

Book 3: Flight Radio for the Private Pilot

The aircraft radio is an amazing piece of equipment. By pressing down on the microphone switch you can have a conversation with a wide range of people; from the local controller to other pilots in the area. But operating the radio also comes with responsibility. To avoid the serious consequences that can result from miscommunication, it is essential that all pilots have a good understanding of flight radio.

Book 4: Principles of Flight for the Private Pilot

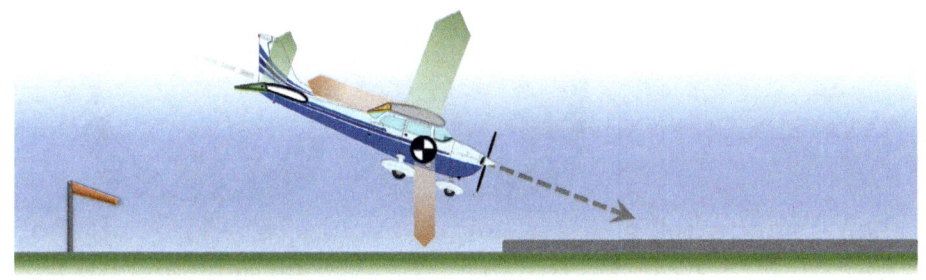

Principles of flight is one of the fundamental topics a pilot must master to operate an aircraft safely. Pilots have so much control over the various forces acting on the aircraft, but flying an aircraft also comes with responsibility. To avoid flying an aircraft beyond its limits, pilots must respect the principles of flight. Turning too sharply, flying too slowly, or overloading an aircraft can all have dire consequences.

Book 5: Flight Navigation for the Private Pilot

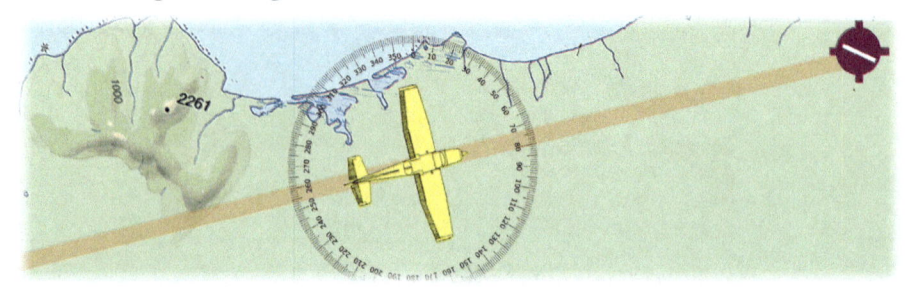

A pilot can travel a considerable distance, across a range of different landscapes on a single flight; from rugged bush, oceans, mountainous terrain and deserts. No other mode of transport offers such freedom. But with this freedom comes responsibility. To arrive safely at a distant destination, pilots must understand the key components of *Flight Navigation.* Flying the wrong heading, underestimating the time and fuel for the flight can all have serious consequences.

Book 7: Flight Instruments for the Private Pilot

When conducting a flight, pilots require a wide range of information. How fast is the aircraft traveling? How high? Is the engine healthy? The job of the flight instruments is to provide all this information to the pilot in a timely and accurate way. It is essential pilots have a solid understanding of the flight instruments in their aircraft, allowing them to obtain critical information, even when they are stretched to their limits.

ABOUT THE AUTHOR

Stephen Walmsley

Stephen Walmsley has been actively involved in aviation for over 20 years. He is a qualified flight instructor, with several thousand hours of flight experience in a range of aircraft. His flying experience has ranged from aerobatics, night flying to multi-engine operations. He holds a PhD in Aviation, with a focus on weather related decision-making.

COMMERCIAL PILOT SERIES

This series builds on the hugely popular *Private Pilot Series*, taking the reader deeper into the world of aviation. Whether you are a new pilot that is exploring each subject in greater detail or a commercial pilot who has been flying for countless years, it is essential to understand the range of factors that can affect a pilot's ability to fly an aircraft safely.

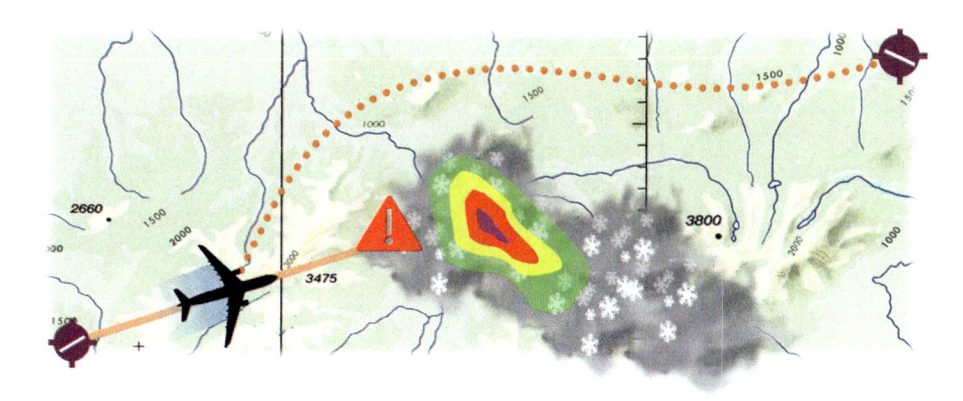

Books in this series include:

Book 8: Pilot Performance & Limitations

Book 9: Meteorology

Book 10: Aerodynamics

Printed in Great Britain
by Amazon